FLINT COUNTRY

A STONE JOURNEY

LAURENCE MITCHELL

Saraband

Published by Saraband
3 Clairmont Gardens
Glasgow, G3 7LW

www.saraband.net

ISBN: 9781916812413

Printed and bound in Great Britain by Clays Ltd, Elcograf S.p.A.

1 3 5 7 9 10 8 6 4 2

FSC
www.fsc.org
MIX
Paper | Supporting responsible forestry
FSC® C018072

The author does not imply that any of the places visited during his travels as described in this book are suitable to visit, either on safety or right-of-access grounds. Please check for hazards or restrictions before departing if you wish to visit any of the locations mentioned. Neither the author nor the publisher can be held responsible for any problem or injury arising at any of the locations in the book.

'An enthralling and beautifully written biography of flint, whose story has been intertwined with humankind's since the very earliest times. Laurence Mitchell invites us to see a world in every knapped flint on the walls of East Anglia's medieval churches, and reveals the practical and magical ways in which flint has shaped our lives and landscapes from prehistory to the present day.' —Fiona Robertson, author of *Stone Lands*

'A good read ... *Flint Country* ventures widely across the various essences that make up the human and natural landscape of East Anglia.' —Stephen Rutt, author of the Wainwright Prize-winning *The Seabird's Cry*

'In his paean to flint Laurence Mitchell celebrates its origins and character as it inspires the work of Neolithic axe makers, mediaeval masons, fire-starters throughout the ages, and the sensuous works of Henry Moore and Barbara Hepworth. In flowing prose we learn how the obduracy of flint has fashioned minds as well as churches, and shaped the history and landscapes of chalklands wherever they surface. A thoroughly enjoyable read.' —David Howe, author of *Extraction to Extinction*

Praise for Laurence Mitchell's previous books:

'Eye-opening and wonderful.' —*Sunday Times*

'A treasure trove of unexpected and little-known facts ... Highly recommended.' —*Outdoor Focus,* journal of the Outdoor Writers and Photographers Guild

'Blends a geographer's store of knowledge with perceptive observations of our natural and man-made environments ... bucketloads of fascinating history and topography ... I felt like I'd been with him every step of the way.' *The Great Outdoors*

'Rich with detail about the landscape ... clear, engaging ... understated ... but, nonetheless, profound.' *Psychogeographic Review*

'A beautifully compelling reminder that we never walk alone. Alongside us, always, are the complex cultural and natural histories, topographies, geologies and folk memories that have helped shape the path and those places we pass.' —Julian Hoffman, author of *Irreplaceable*

For Jackie

The great affair is to move; to feel the needs and hitches of our life more nearly; to come down off this feather-bed of civilization, and find the globe granite underfoot and strewn with cutting flints.

—Robert Louis Stevenson

CONTENTS

INTRODUCTION

Our house is filled with stones. Not overflowing exactly, but an assortment of pebbles and small pieces of rock inhabit most of the rooms one way or another. Placed along bookshelves and window ledges, or piled in dishes like miniature cairns, they are the product of absent-minded collection over the years and represent episodes of foreign travel, trips to the beach and a more general magpie-like hoarding of the bright and shiny. Unlabelled, their exact origin has been forgotten in many instances. Gathered with a traveller's impulse, the stones serve as souvenirs of place, mementos of journeys to exotic locations.

A few are so memorable in regards to time and place that I can recall the precise circumstances in which they came into my possession. There is a piece of dusty mauve amethyst bought for a few dirhams from a child somewhere in the Atlas Mountains on my very first trip outside Europe. Then there are the mineralised bat droppings like tiny meteorites that I gathered from a stream deep inside a Sumatran cave. Elsewhere in this miscellany I have stones picked up from Egypt's White Desert, pieces of sponge-like volcanic tufa from an Indonesian volcano, rocks gathered from Iceland's Þingvellir valley where the North American and Eurasian plates collide, pieces of two-billion-year-old Lewisian gneiss from the Outer Hebrides, and lumps of pink granite from who knows where – Cornwall? The problem is that, while I gather and collect, I do not curate with anywhere near the same enthusiasm and so the provenance of many of these stones is a mystery to me. Most have found a place in our home for reasons of shape, markings or colour; few have been collected with anything approaching a geologist's eye.

As for flint, there is plenty of that lying about in the house too, gathered from around the British Isles, many from Norfolk and Suffolk beaches, at one time or another – wave-smoothed nodules that have been chipped in places to reveal the shiny black (or grey, or brown) treasure within. There are a few sharp-edged fragments too – 'microliths', as archaeologists like to call them – picked up on the chance that they might once have been held by the hand of someone who lived and died an unimaginably long time ago.

The haphazard collection of such things is an ongoing process. As the words slowly gather writing all this down, so do the flints that litter my desk.

ONE

FLINTLOCKED

Norwich – Norfolk

Deep assignments run through all our lives; there are no coincidences.

J. G. Ballard *The Atrocity Exhibition*

For years I harboured a mild obsession with stones that were in the wrong place. It was about those rocks that had strayed across the geological map and were no longer where they were supposed to be. In my case it was the glacial erratics of East Anglia: those ice-hauled aliens from the north – chunks of hard rock from Yorkshire, Scotland and Norway that had been prised away from their native deposits and transported by slow but unstoppable, mile-deep sheets of ice to the low country of eastern England. Too heavy and ungainly to budge from where they eventually came to rest, the stones remained in situ as the ice melted and sea levels rose to render their accidental homeland an island once more. Millennia passed as they lay like thrones upon the land, unmoved in their new habitat, dormant in the hedgerows, fields and ditches that would later come to define the landscape. In time, local legend would adhere to these stones like moss and lichen – folk tales of sleeping giants and ogres, of magical powers, of the floods and disasters that would ensue if anyone were foolish enough to attempt to move them. Even so, a few were manhandled short distances to rest in churchyards or to lay upon village greens where people could keep an eye on them and even sometimes use them as a centrepiece for gatherings and rituals.

The Anglian Ice Sheet that covered much of Britain nearly half a million years ago extended as far south as Essex. With the ice came boulders that had been prised from their native bedrock and shifted great distances south by the slow moving glacial action. Many of the northern rocks transported by the glaciers remain where they were deposited to this day. Hard rock loners in a soft southern landscape, they are geological upstarts: alien scatterings on the fields, strangers on the shore.

Norfolk, where I live, has a good number of glacial erratics scattered within its borders. There is no particular reason why this county should be particularly favoured: it is simply that they are more conspicuous here in a flat, largely agricultural landscape. Who, other than a hawk-eyed geologist, takes notice of a wayward rock in a craggy, stone-littered landscape like the Lake District? Here in East Anglia, a region dominated by Quaternary deposits over deep-seated chalk, such rocks are curiosities, objects of wonder and myth.

There is another rock, though, equally conspicuous, that is commonplace and native to this region. It lies buried under the chalk and scattered on the fields and beaches but is so much part of the scenery that its ubiquity is easy to overlook. Unlike wayward rocks from the north, flint is indigenous, it belongs here. The touchstone of the East Anglia landscape, flint beckons us to notice. Sometimes, though, we just do not see it.

*

The Covid-19 pandemic that spread across the world like a malevolent miasma in the spring of 2020 stopped all of us in its tracks. After weeks of speculation and uncertainty normal life was abruptly frozen in place and put temporarily on hold. As temporarily as possible was the hope but no one could say for how long; nor wanted to, other than hesitate to state that it would, like the Great War, probably be 'all over by Christmas'. The passing of time in such circumstances takes on a different character when the endpoint is unknown. This was unknown territory. The only meaningful reference points were long-elapsed events whose folk memory still engendered an existential fear: the 'Spanish flu' epidemic of 1918 that infected nearly a third of the world population and killed more individuals than the world war it succeeded; or going back further, the Black Death that raged between 1348–53 and later resurfaced with the Great Plague of London in 1665–66.

Almost overnight, movement was restricted, human contact was to be feared and we were confined to our houses in an attempt to control the spread of the mysterious although clearly deadly disease. The first lockdown took place, somewhat contrarily, during a glorious spring of fast-rising sap, balmy weather and perpetual blue skies. Life was tolerable enough as long as you had a reliable supply of food on order and some sort of access to an outside space. For some – those more introverted among us, perhaps – the sudden change in everyday life came almost as a relief: the modern world had stopped spinning for a while and despite all the fear and uncertainty, all the unconvincing press conferences and sometimes contradictory information on how best to deal with this new viral menace, it was an opportunity to slow down and take store. It became a time to reflect, a time to breathe (although an inability to breathe was, at this time of oxygen-robbing pestilence, perhaps the greatest fear of all).

For some the enforced home time was a chance to be creative, to develop new skills – a foreign language maybe – or improve nascent baking talent. Others seized the opportunity to make inroads on the supermarket merlot a little earlier in the day than they might normally do. This is not to make light of it; for many, particularly the sick, the elderly, the poor and vulnerable, or young families trapped in small high-rise flats with children, it was a time of hardship fraught with economic and psychological difficulties. Time not so much stood still as circulated slowly with tedious repetition – for many, these were grim groundhog days that had to be endured rather than savoured. Those of us fortunate enough to be not affected by illness or privation continued as best we could, embracing new routines, making our daily walks, reading, writing, cooking, gardening – things we routinely did anyway but now more self-consciously with a little more purpose and focus.

During this enforced period of immobility and social exclusion it was widely noted that people were starting to pay more attention to things that had previously gone unobserved. The circumstances were certainly favourable. For one thing, the urban soundscape had changed. The lack of traffic and commercial activity suddenly made the world – certainly the city environment – a quieter place.

Many of us took refuge in our gardens, if we were fortunate enough to have one, or in our government-sanctioned exercise time – an hour's daily walking from our homes. As someone who works from home and who requires not only the exercise but also a change of scenery to prevent the onset of cabin fever this was a habit I had already long subscribed to. I walked every day, either on my own or with Jackie, my wife. It was a daily highlight to step out into almost traffic-free streets beneath an aeroplane-free sky, the familiar lo-fi hum of the city's constant mechanical kinesis now reduced to little more than a whisper. It was surreal, sometimes unsettling, to see normally busy roads almost completely car-free. There was something almost apocalyptic about it, something reminiscent of the opening scene of Danny Boyle's *28 Days Later* in which Jim, the coma-woken bicycle courier played by Cillian Murphy, staggers bewildered through the streets of an eerie and menacingly deserted London. It was strange, too, to contemplate where everyone had gone. They were, of course, simply at home. Every house was occupied, couples and families locked down together in terraces, flats and bedsits, all waiting for news as they watched the televised briefings each afternoon and contemplated what the future might bring in this strange new world.

As for walks, there were several options. A short drive in the car meant that I could take a turn around one of several places at the city's edge – a wildlife area of marshland and grazing next to the river, or a circuit of a lake that had been created by the flooding of an extensive gravel pit – but the rigours of strict 'lockdown' encouraged a new Puritanism and for a while we left the car in the garage and simply walked from our front door. Again, there were modest choices to be made and over time a number of regular circuits fell into place, walks that registered both in our minds and in the muscle memory of our legs. These mostly involved a grid of pleasant, if dull, suburban streets but they also incorporated a couple of city parks, an old railway line that has been converted into a pedestrian and cycle route, and sometimes a church or two – Norwich has them in spades, or more accurately, in dozens. Sometimes the walk led down to the river that flows through this city. This would give us a tenuous connection with the sea, as even this far inland the river is still noticeably tidal and if the wind is blowing in the right direction it seems possible to detect a whiff of the briny North Sea that lies twenty miles beyond.

One regular walking circuit led along a disused railway track to a suburb of Victorian terraces. Central to this area, a little way south of the city centre, is a large turf-covered water works, a recreation ground with a bowling green and a church. The tower of the church with its four corner turrets is a landmark in the area and more often than not I found myself heading through the church's gently solemn, ivy-tangled graveyard.

St Mark's, which stands central to the tree-shaded rectangle of the churchyard, like much of the surrounding housing, is of Victorian vintage, built in the 1840s in Anglican Revival style. The church was invariably locked and so there was no opportunity to go inside and admire its box pews and unusual rood loft and screen.[1] It was St Mark's exterior that captured my attention though. In an old city like Norwich, a city replete with far more than its fair share of glorious medieval churches, St Mark's might be considered unremarkable. It is, after all, relatively modest and modern: no great antiquity or breathtaking proportions, no Norman arches or towering Perpendicular windows to consider; no fine stained glass (its original windows had been blown out by the Baedeker air raids of the Norwich Blitz in 1942). But there was one thing that this Victorian upstart had in common with the thirty-one surviving medieval churches that were clutched like precious eggs within the nest of the city walls that lay just a five minute walk to the north: the material that it was largely built from – flint.

Constructed primarily of knapped flint blocks, on sunny days the walls and tower of the church glimmered with the sunlight that filtered through the churchyard's leafy canopy. In cloudier weather, rather than glistening, the walls glowered with deep intensity, the flint's shiny surface catching the light with same ambiguity as a starling's feathers. Although still a beacon for the neighbourhood it stood in, the harsh reality was that currently the church was a building that could not fulfil the purpose for which it had been originally constructed. Lockdown had sealed its doors; its altar and pews were redundant, bells silent. The church stood, like so many others, an empty repository of service and celebration.

With its knapped flint walls and Victorian Gothic style, St Mark's was a nod to the medieval churches in the city centre; a salute to the medieval masons who had constructed these churches with little more than basic tools and precarious wooden scaffolding. The church's flint work, although extensive, was relatively modest. The flint blocks that made up most of the walls and tower had been worked by hand to show one shiny face to the world. Rounded in profile and of similar size, they had been arranged into approximate rows set in mortar. The impression was rustic, East Anglia vernacular – these might be flints gathered from fields or shingle beaches on the coast. In contrast, the flints used to pattern the narrow buttresses that supported the external walls had been more painstakingly knapped into squares that fitted snugly together. As decorative as they were structural, they gave the church an elegance that brick alone could not have achieved. Even so, compared to some of the exquisitely worked flint that ennobles many of the churches and civic buildings in the city's historic core, the work here was more utilitarian than inspired. The flint was dark but by no means uniform. I fancied that the stone used on the north side of the church was more brownish in tone than that on the opposite side, which to me appeared more greyish. Observing the tower from some distance

down the driveway, I also thought that the flint placed higher up on the tower looked lighter and more silvery. Was this deliberate? Had the choicest pieces been set aside for positioning high up on the tower to reflect the light and act like a beacon?

On some visits I got up close to the stone itself. Each square block was markedly different from its neighbour. The colour varied, as did the contours of the surface that had been knapped to produce a relatively flat face to line up with the others that combined to make the wall. Examined close up, every flint was a world in miniature, a fluid silicon sea solidified into hard rock. Each square face was like an abstract painting, a cell in a composite art work. Some, I noticed, contained tiny fossils: once living things that were older than the rock that held them; stranded like rafts in a sea of stone, each one with a story to tell. The church's flint held time captive, a snapshot of a precise point in geological history, although it was a timepiece that I didn't know how to read.

After several visits to the church, looking at the detail of it more attentively each time, a question arose in my mind and kept resurfacing: where did all this flint come from? The immediate answer was obvious – it was locally sourced – but only then did it really dawn on me that in all my years in the city I had been taking the ubiquity of flint completely for granted. As an outsider to the region, albeit one who had lived in Norfolk for most of his adult life, I should have known better; I should have noticed. As someone who had grown up in a part of the country where the rocks were much older than here, flint was still a novelty. The Midlands where I came from were not part of that world. No chalk, no pebbles, no beaches – indeed, no sea – just rich red sandstone in an undulating landscape. When I made the move east as a young adult what I noticed most about my new environment was nothing to do with the bedrock, sandstone, chalk or otherwise, it was simply an absence of hills. I still miss the hills.

The landscapes in which I now spend most of my days – the urban environment of a medieval city and the flat, largely arable countryside that surrounds it – remain foreign to the one that is imprinted in me. The markers of early life are hard to shake off. We cling to the same values that were formed in infancy, the same tastes for music and art that we acquired in adolescence; the same appreciation of landscape – and if not exactly 'appreciation' then a certain ease of living within it. The landscape imprints itself in more ways than one. Archaeologists make use of oxygen and strontium isotopes in teeth to pinpoint where someone grew up. Teeth formed in childhood carry the chemical signature of the water and plant food taken by the individual. This becomes fixed in tooth enamel and remains for life. Our teeth are tied to a particular geology – that of the rocks that our childhood drinking water percolated through.[2] Like many twentieth-century Midlanders, my teeth are Welsh; more specifically, they bear the geological signature of the reservoirs of the Elan Valley in central Wales, where the hard Lower Silurian sediments are 420 million years.

We may move away to broaden horizons but in our hearts we cling to the landscape of childhood however mundane that might be. I grew up in north Worcestershire – the green belt suburbia of a Birmingham overspill New Town, a semi-rural edgeland of leafy lanes that lead to rolling pasture: sheep, cows, bucolic vistas, the distant undulations of the Lickey and Malvern Hills. Later in life I would seek out exaggerated versions of these hills, which I would experience through the lens of my imprinted landscape – the Scottish Highlands, the Peak District and Lake District. Later still, I would be drawn to even more dramatic uplands like the Caucasus and Himalayas, those thrillingly exotic high places with their cloud-piercing peaks.

Norfolk is different – contrarily horizontal – which is perhaps why it took so long. Places take time to sink in, to become represented in the consciousness, to become mythic. The imprinted landscape lingers in the mind and provides a filter through which any new environment is interpreted. As the Midlands poet Roy Fisher once casually but meaningfully remarked: 'Birmingham is what I think with.' My own primal landscape, the territory I carry with me internally, is not so far off geographically speaking. It may be that my fascination with glacial erratics, in fragments of displaced upland rock in a lowland landscape, is also somehow connected to this.

There is a type of cognitive bias known as the Baader-Meinhof phenomenon that describes how after first becoming aware of something it can seem as if you encounter it with almost uncanny regularity. A simple explanation for this phenomenon is that which can seem like a strange coincidence is really no more than a matter of frequency illusion – that which previously went largely unseen to the conscious mind suddenly becomes something that is encountered far and wide. It is a common enough experience but it can feel disconcerting, almost as if you are under some sort of spell or agency directing a particular thing into your gaze.

This is what happened to me regarding my newfound interest in flint. After pondering what I saw at St Mark's I seemed to come across the stone everywhere I looked. This, you might argue, is hardly surprising in somewhere like Norwich, a city of medieval buildings in the heart of one of the flintiest parts of the country. The thing is: now I had started to pay attention to it. Something connected with the mineral influence of a familiar landscape, could it be that flint was offering an answer to a question that I didn't even realise I was asking?

In Norwich, flint is a fundamental component of the urban landscape, part of the fabric of the city. My home lies just outside the old city walls, close to one of the former city gates. The gate is long gone but sizeable fragments of the original fourteenth-century city walls and its towers still stand along the route of the inner ring road. The walls were built largely of flint – great dark lumps of it held together with rough mortar and odd patches of brick and stone reinforcement. Flint was also the principal building material used for many of the churches and civic buildings in the city centre. But it wasn't just here in buildings and walls, it was ubiquitous – in garden rockeries, on doorsteps, embedded in the cobbled service roads that ran between the terraced streets of the inner city. On one of my walks I even came across a private road in the southern suburbs where a succession of large flint nodules had been spaced out along the grass verge to discourage parking.

While it was tempting to attribute my newly acquired awareness to cognitive bias alone, during that first Covid spring there were coincidences involving flint that stood out like writing on the wall. One occasion was precisely that, although it was a matter of drawings rather than writing per se. One day wandering through the narrow lanes that lie close to the marketplace I came across a white-painted gable that bore a mural of Stone Age flint axes and arrow heads. The stone tools were rendered as line drawings and looked as if they had been copied from some archaeological textbook. The place itself was shuttered and locked, Covid-closed for the duration. A sign by the door bore the legend: *FLINT*. Other than this there were no further clues as to what went on within. At first I wondered if it was something like an emporium for reproduction stone tools, although that seemed too niche even for a city like Norwich with its admirably broad spectrum of particular and sometimes peculiar obsessions. Back home, a Google search revealed the business to be a women's hairdressers ('Old and New Styles'). It was nothing more than a name coupled with a pleasing mural, and I could only assume that, rather than flint blades, combs, curlers and scissors were the tools employed in the cutting business here, even for the 'old styles'.

The hairdressers' mural was another reminder about the stone that had recently seeped its way into my consciousness. Long before flint was employed for building or road-making, and even before mankind lived in anything that resembled a building, the stone had been a precious resource. At some time in the distant past hominin ancestors had taken note of how flint yielded a razor-like edge when fractured: an edge capable of skinning and dismembering an animal; an edge keen enough to cut through flesh, and if attached to some sort of projectile sharp enough to penetrate the hide of the animals that provided them with sustenance. Such sharp edges were easily created and, with a modicum of practice, efficient and highly portable tools could be made. Portability was a vital element as this seemingly miraculous stone could not be found everywhere.

At some other time in the same distant past it also became knowledge that, struck hard, this same stone could generate sparks and thus provide the gift of fire. Like its facility for providing blades, the discovery of flint's fire-starting potential likely took place around the globe at approximately the same time: a matter of synchronous discovery that resulted in parallel human development. It was such discoveries that propelled human evolution forward more rapidly than anything else that had come before it.

The Industrial Revolution that began in the English West Midlands in the late eighteenth century turned the world upside down in a matter of a few decades – a technological surge forward that brought unstoppable progress as well as, in hindsight, irreparable damage. But long before any of this came about a far more fundamental revolution had already taken place. This was the advancement of technology and material culture that came with the development of stone tools by early hominins. With tools made of flint and similar stone humanity was able to progress at its own slow pace, to eventually develop the social systems and political frameworks necessary for advancement. It was the use of flint that set the chain of events in motion: a precursor to a series of technological leaps and bounds that eventually gained so much momentum that humanity would trip itself up along the way.

*

The idea for this book came quickly; or, dare I say it, as a spark of thought. Its realisation was far more glacial in pace – a slow burn (to flagrantly mix metaphors), a gradual smudging of the page. Naturally enough, this is all about flint in one way or another, its place in the landscape and in human cultural history. Effectively, this is a journey into the landscapes of a singular type of stone, an expedition through deeply folded time and layered geology. A stone journey: a pilgrimage of sorts.

Religious pilgrimages sometimes involve the use of stones. In the Christian faith, stones are used as way markers along the route of the Way of St James to Santiago de Compostela in northern Spain. In Islam, at a specific point in the Muslim Hajj, stones are picked up and thrown at a wall as a symbolic means of driving away the Devil. At the climax of the journey in Mecca, pilgrims circulate around the black cube of the Kaaba at the centre of the courtyard of Masjid al-Haram, Islam's holiest mosque. Set into a corner of this structure, the most sacred site in Islam, is the Black Stone, *al-Hajar al-Aswad*, which is believed to have been put in place by the Prophet Mohammed himself. Although it actually dates from pagan times, the stone is revered by Muslims and considered to be an Islamic relic. According to tradition the stone, which is highly polished by the repeated touch of an untold number of pilgrim hands, dates back to the time of Adam and Eve when it fell from paradise to indicate the place where the couple should build an altar. Another tradition states that the stone was formerly an angel who was turned into a jewel as punishment for failing to warn Adam about the apple in the Garden of Eden. Consisting of not one but several stones cemented together in a silver frame, the Black Stone is widely regarded as being a meteorite although there are others who think that it could be basalt, agate or obsidian. It might therefore have an origin that lies either in deepest outer space or as molten rock flowing deep beneath the Earth's crust. Either way, extraterrestrial or subterranean, it is heavily laden with meaning and metaphor.

Flint, though, is born of the sea, albeit a warmer sea at a time when the world looked very different to how it does today. We are all born of the sea if you go back far enough.

TWO
FLINT EASTWARD

Norwich, Burgh Castle, Caister St Edmunds – Norfolk

I feel that I have seen enough of flint houses, flint chapels, flint sheds, flint garden walls and heaps of chalk-whitened flint to last me a lifetime.
William A. Dutt *Highways & Byways in East Anglia* (1901)

Some landscapes wear their heart on their sleeves. Their topography tells the whole story: hills and mountains, deep valleys, craggy vistas, rocky outcrops – geomorphology writ large, geology in plain sight. To understand the low country of East Anglia, though, you need to dig down to what lies beneath the surface. Underground, beneath the hedges and ploughed fields, the village streets, the car parks and out of town shopping centres, lies a layer of chalk up to half a kilometre thick. In places the chalk is deeply submerged beneath deposits of sand, clay and gravel;

elsewhere it lies closer to the surface, or even on the surface itself – its telltale whiteness beckoning the eye in a landscape that is predominantly green and brown.

Chalk is the mother of flint or, rather, it is the womb from which it is born. Although secondary flint is sometimes found at great distance from its origin, it is from chalk that it comes into the world in the first place. Companions in the same rock stratum, chalk and flint could not be more different, representing opposite ends of the mineral spectrum. While chalk is relatively soft, crumbly and white, flint is hard, almost without grain, and often dark in colour. Given these extremes, it is tempting to say that the two are like 'chalk and cheese', although the phrase seems a little weak as notwithstanding its inedibility chalk is, texturally speaking, actually quite like certain varieties of cheese – Lancashire or Wensleydale, say. Flint, though, is as it sounds – hard, unyielding, sharp-edged, yet brittle.

Chalk does not so much bear fossils as *is* fossil: uncountable billions of the calcified shells of microscopic sea creatures that lived around one hundred million years ago. The time of the chalk, the Cretaceous period, was the last of the three periods of the Mesozoic era, an era that lasted for the best part of 200 million years and began and concluded with cataclysmic mass extinction events. It ended abruptly around 66 million years ago with the Cretaceous–Paleogene (K–Pg) extinction event that saw the rapid demise of all non-avian dinosaurs, and those distinctive creatures that we generally associate with the limestone rocks of the Jurassic period that preceded the Cretaceous – the coiled, supersized shell dwellers that we know as ammonites.

Buried deep beneath the earth or gleaming white as it breaks the surface, chalk underlies a large portion of southern and eastern England. One of England's most fêted landscapes is that of the chalk cliffs of the southeast coast – the famous White Cliffs of Dover, which have long served as a totem of national identity and island nation sovereignty. Like all landscapes, Dover's white cliffs represent a place where matter and metaphor converge, where genius loci and geology coalesce into meaning. A vast wall of pristine chalk cliffs that faces imperiously across the English Channel towards France, Vera Lynn sang nostalgically about this iconic landform during the darkest days of World War II. The song '(There'll Be Bluebirds Over) The White Cliffs of Dover' immediately became an anthem of spirited British defiance to the Nazi threat that loomed just across the channel in mainland Europe. It is rare for an English landscape feature to be celebrated in such a way but it is a song written by Americans who, as a rule, are less bashful about such things. For pedants though, the song is irretrievably flawed: its lyricist, Nat Burton, while familiar with southern England's geology, knew little of its ornithology and did not realise that bluebirds belong to a North American genus that is entirely absent from Britain.

Chalk is more or less pure calcium carbonate. The chalk of the Dover cliffs is the same stuff as that used on blackboards in the school classrooms of yesteryear. Powdered chalk has non-slip properties that come in useful for dusting the end of snooker cues or rock climbers' fingers. The same powdered rock is also used in anti-indigestion remedies or for sweetening acid soil – chalk is alkaline and reacts with acids to neutralise them.

Chalk's whiteness has long been valued by those who have lived in these landscapes. The supersized horse figures inscribed on chalk hillsides in southern England, and the boldly priapic figure of Dorset's Cerne Abbas Giant, owe their visibility to the rock that has been incised beneath. The white chalk gleams and stands out from the grass cover, delineating the figures clearly in the landscape. Chalk has been used to enhance other celebrated manmade structures too. The volcano-like mound of Silbury Hill in Wiltshire near Avebury may be astonishing enough as it stands but it should be remembered that when it was first constructed, before a thin layer of soil built up and grasses had chance to take root, it was a pure white chalk structure that would gleam like a beacon in the sun (as did the Pyramids of Giza, which were originally covered in blocks of polished limestone).

It is chalk that characterises much of the landscape of the southern counties of Sussex, Kent, Hampshire, Wiltshire and Dorset, the Chiltern Hills, the Isle of Wight and the Lincolnshire and Yorkshire Wolds. These Upper Cretaceous formations are not unique to England but continue beneath the English Channel and North Sea to resurface in France, Germany and the Low Countries. In Britain, away from southern and eastern England, exposed chalk becomes scarce but similar chalk formations can also be found on Northern Ireland's Antrim coast, along with a few scattered outcrops on the Inner Hebrides and western mainland of Scotland.

In East Anglia, a thick band of chalk lies beneath the surface all the way from The Wash to the North Sea coast of Norfolk and Suffolk. It is not level but a tilted sheet that dips south-east towards the North Sea. Laid down in shallow warm seas between 135 and 70 million years ago during the Cretaceous period, the chalk is up to 450 metres thick and underlies the whole of Norfolk east of the escarpment of older rocks that mark the boundary of the Fens. At Great Yarmouth on Norfolk's east coast the top of the chalk is 150 metres below ground level; in the west of the county it lies much closer to the surface.[1] The chalk is overlain with far more recent glacial deposits. The Devensian, the last glacial advance across northern Europe, which lasted until about 10,000 years ago, stopped shy of Norfolk but the county's surface geology of Pleistocene crags, clays, sands and gravels confirms the widespread effect of ice on the landscape over many millennia before this.

While eastern England may have some of the youngest rocks in Britain, it is wrong to think that the region sits exclusively on relatively recent geological foundations. It is really a matter of longitude and depth: a bore hole sunk in northwest Norfolk in 1946 revealed a 500 metre-thick layer of 200 million-year-old

Triassic rock deep beneath the surface. Beneath this, starting at a depth of 800 metres, lies a pavement of pre-Cambrian rocks that are more than 600 million years old.[2] Simply put: the further west and the deeper down you go, the older it gets. Deep time lies deeply buried; geology is its chronometer.

Although it originates in layers in chalk, flint is often found at considerable distance from its source. The pebble beaches found in East Anglia and along the south coast are testament to this. These exist largely as a result of longshore drift, a dynamic process by which sand and stones are moved by currents that run parallel to coastlines. On the southern English coast, the beach at Brighton and the great shingle 'nose' at Dungeness in Kent are composed almost entirely of flint pebbles. This is hardly surprising given the proximity of so much exposed chalk but even away in areas where different rock predominates, such as Chesil Beach on Dorset's limestone-rich Jurassic Coast, nine-tenths of the beach pebbles are also of flint, having originated in chalk elsewhere and been transported here by longshore drift.[3]

Like the northern erratics dispersed by glaciers, many of the flints found on the shoreline, as well as those found scattered on the fields, have been shifted far from their place of origin. These scattered rocks and pebbles are what geologists call Quaternary flint: stones subjected to movement during the age of the great ice sheets over the past 2.6 million years. Many have found their place in this world during the most recent epoch of the Quaternary, the Holocene that began 11,700 years ago, the same epoch in which human development fast-forwarded from Stone Age hunter gathering to sophisticated urban living and nuclear technology. Geology, like water, like humanity, is in a perpetual state of flux. As the writer and cartographer Tim Robinson has noted, 'the geographies over which we are so suicidally passionate are… fleeting expressions of the earth's face.'[4]

*

The Norfolk coastline is a mixture of mudflats, marshes and wide sandy beaches but there are long stretches of Quaternary flint shingle on the north coast at Cley next the Sea and along the spit at Blakeney Point. Much of Suffolk's coastline is pebble or shingle too, with the onetime top secret military installation of Orford Ness being located, away from prying eyes, on a shingle spit so remote that it is effectively an island. The aptly named Shingle Street, an isolated coastal settlement south of Orford Ness, has its own secret wartime associations too.

Writing at the turn of the twentieth century, the Suffolk writer and naturalist William Dutt claimed to have seen so much flint in his rambles around East Anglia that he came to disdain the stone. Little wonder then that in later life he turned his energies to documenting glacial erratics, or rather what he termed 'mark-stones', in his 1926 pamphlet *The Ancient Mark-stones of East Anglia*. As I outlined earlier, it was these curious aliens in the landscape that eventually led me towards an interest in the region's more native geology.

With so much of it lying about on the fields, on the seashore and embedded in the chalk below, it is only natural that flint has long been exploited in those areas where it is plentiful.

Flint has long been used as a building material across Norfolk, Suffolk and parts of Essex and Cambridgeshire. In Norfolk, the seafaring towns of Great Yarmouth and King's Lynn that bookend the county's coastline both have impressive flint-built buildings. Yarmouth's town walls, far better preserved than those in Norwich, are also constructed largely of flint. King's Lynn, a one-time Hanseatic port, has no walls but instead possesses a splendid fifteenth-century Town Hall with elaborate and eye-catching chequerboard flint flushwork. Elsewhere on the Norfolk coast, in towns like Cromer, Sheringham and Wells-next-the-Sea, and in gentrified one-time fishing ports like Cley next the Sea and Blakeney, domestic dwellings constructed of flint beach pebbles can be seen on almost any street, and in the north Norfolk hinterland flint pebble barns and farm buildings dating from the late eighteenth and nineteenth centuries are also a common sight. Away from the coast, in the Brecks region that straddles the Norfolk–Suffolk border, the small towns of Thetford and Brandon also have a wealth of flint buildings, the latter being associated with an extraction industry that persisted into the twentieth century. And beyond this secular realm, magnificent medieval churches can be found in every corner of the region, either standing alone or as the centrepiece of a village green, their flinty towers raised in tribute to the broad East Anglia sky.

*

Visitors to Norwich have long noted the abundance and splendour of its flint buildings. The equestrian traveller Celia Fiennes visiting the city in 1698 observed that Norwich, in addition to having 'a great number of dissenters' was 'a rich, thriving industrious place'. She also remarked on the large number of churches in the city:

> *They are built all of flints well headed or cut which makes them look blackish and shining.*

Fiennes was equally taken with some of the domestic architecture she found:

> ... *by one of the churches there is a wall made of flints that is headed very finely and cut so exactly square and even to shut in one to another that the whole wall is made without cement at all they say, and there appears to be very little, if any, mortar; it looks well, very smooth shining and black.*[5]

The building whose wall Celia Fiennes was so impressed with still stands and for almost a century has served as the city's

Bridewell Museum. The wall next to the museum entrance bears a plaque with a coat of arms that reads:

> *The Arms of WILLIAM APPLEYARD, first Mayor of Norwich 1403-4 to 1405 who served his Mayoralty in the adjacent house built in the latter part of the XIVth Century and esteemed the finest piece of flintwork in England.*

Built in 1325 by a wealthy merchant called Geoffrey de Salle, the property was enlarged in 1386 by William Appleyard, who would later become the first Mayor of Norwich. In the sixteenth century, another Mayor, William Gardner, took possession of the property for a while until it was sold by John Sotherton to the Corporation of Norwich in 1580. From this time forward the building would have a civic function in the city. Its first public role was as a house of correction where the city's beggars and hapless poor were put to work, encouraged in their labour by 'corrective' means such as whips, shackles and stocks. The employment involved physical toil that was both arduous and compulsory; men were given manual jobs like wood turning to perform, while women were set to work spinning and carding wool. It was around this same time that William Appleyard's house became known as the Bridewell in reference to the London prison of the same name. The name Bridewell later became synonymous with larger prisons, and in Norwich it is a name that continues to this day even though the building now provides a far gentler form of personal improvement. The Bridewell remained a prison until the early nineteenth century, after which it found alternative use as a tobacco factory, warehouse and finally a shoe factory, one of Victorian Norwich's boom industries. In 1928 it became a museum. Although much of the original building has been destroyed and rebuilt over the years, the long external wall of tightly fitted, black flint squares remains much the same as it was when it stood gleaming new in the fourteenth century.

The Bridewell's extensive flint wall faces onto a narrow alleyway, on the other side of which stands the south wall of St Andrew's, a chunky-towered flint church in Perpendicular style that unlike many of the city's other medieval churches remains a working place of worship. Built between 1470 and the start of the 1500s, its walls are a century older than those of the Bridewell opposite and its masons must have had the older black flint wall in mind when they were building the church. This is perhaps what makes the Bridewell so unusual: at a time when much of the wealth from the wool trade was being piously invested in ecclesiastical masonry a similar-sized purse was being spent on what was effectively a private residence. A forerunner of some of the grand flint-worked civic buildings that would be commissioned in later centuries, the Bridewell demonstrates one of the earliest, and certainly most splendid, examples of skilled flintwork in a non-religious building in the country. As the plaque by the museum entrance confirms, it has long been considered *the finest piece of flintwork in England*.

It was perhaps in the spirit of the Bridewell that one of Norwich's most attractive public conveniences was built. Just down the alleyway from the Bridewell, on the corner of St Andrew's Plain and Princes Street, is an elegant round-fronted building of neatly squared black flint that echoes both St Andrew's Church and the nearby fourteenth-century wall. Built in 1892, in a style that was intended to blend in with the surrounding architecture, the building was constructed as part of a well-meaning Victorian campaign to make the streets of inner city Norwich urine-free (perhaps a thankless task in a city with over 400 pubs). Sadly, it has been boarded-up for decades and its interior porcelain, alas, no longer sees any active service, although many of Norwich's male citizens, myself included, can remember it serving as a conveniently located gents for many years.

Norwich's other main civic flint building par excellence is the Grade I listed Guildhall that stands close to the marketplace.

Like the Bridewell, this has seen several changes of use during its lifetime. It was built as a civic meeting place in the early fifteenth century on the occasion of Henry IV awarding Norwich autonomy from the county of Norfolk in 1404. In 1511 the roof fell in and subsequent rebuilding was completed by 1535. Like many of the splendidly ornate churches of the same period, the Guildhall's elaborate design is a clear reflection of the prosperity enjoyed by the city's ruling elite as a result of the medieval wool trade. Considered to be the most elaborate of England's provincial medieval city halls, at various times it has served as a prison and law courts, the same functions as the toll house that had previously stood on the same site.[6] Thought to have been inspired by grand buildings in the Low Countries, with which Norwich was well connected, the Guildhall multi-purposed for many years as the city's assembly chamber, sheriff's courts, armoury and gaol. Britain's first Protestant martyr, Thomas Bilney, was once incarcerated in the building's undercroft before his execution. The Guildhall continued as the seat of city governance until 1938 when the modernist City Hall was built at the top of the marketplace. Since then it has seen service as offices for small businesses, as a tearoom and as a booking office for the annual Norfolk & Norwich Festival. The Sheriff's office remains in place in what was once the Guildhall chapel.[7]

The Guildhall's finest feature is undoubtedly the flint flushwork on the upper part of its east wall, the exquisite workmanship of which is clear to even the most untrained eye. Consisting of interlocked diamonds of black flint and white limestone, the choice of chequerboard patterning is possibly a reflection of the Guildhall's function as an exchequer, where a squared cloth was traditionally used for counting money. The lower course of the wall is of square-knapped black flint that appears to be held together with a minimum, or even an absence, of mortar. This is similar to the flint work on the Bridewell but is actually superior in having precisely level courses of uniform sized knapped flint

rather than more randomly fitting squares. Elsewhere, like on the later porch extension on the south wall, small shards of flint has been used to decorate the mortar between more roughly shaped flints – a technique known as galleting. Constructed, as the website of British Listed Buildings asserts, using 'Quaternary and Quarry Flint with Lincolnshire Limestone',[8] both the quality of the materials and the consummate skill employed by the masons are plain to see. But, if the limestone for the ashlar masonry was brought in from Lincolnshire, the question remains: where was the flint obtained and how local was it? 'Quaternary' here refers to what elsewhere is sometimes called 'derived': flint that has been eroded into pebbles through transportation from its original home in chalk by way of ice sheets, rivers and ocean currents. 'Quarry Flint', of course, speaks for itself yet the precise location of the quarry in question is often unknown. Such details tend to go unrecorded.

Who were these masons? Where did they learn their skills? What was their sense of the place their own lives would have in the historical record? And what was their sense of time? Did they ever consider that the building they so painstakingly worked on would still be in use half a millennium later? As for the high quality flint they used, they must have considered the shiny black stone they worked with as simply a gift from God. It was a material with which they were highly adept, a stone they could skilfully manipulate to make the most of its inherent grain-less beauty. The masons had no notion that the raw stone they held in their hands had once been something organic and liquid; a stone composed of the hardened silica gunk that had oozed out of microscopic sea creatures an unimaginably, and un-biblically, long time before. How could they? According to the religious orthodoxy that prevailed at the time, the world was no more than a few thousand years old.

*

The adoption of flint as a building material did not begin with the fevered period of medieval church construction that followed in the wake of the Norman Conquest. While the medieval period may represent the climax of flint masonry, the stone had been used as a building material in the region long before this. Although the majority of Saxon churches were constructed of wood and have not survived, some flint churches, in particular those with round towers found so commonly in Norfolk, are likely to be of pre-Norman origin. It was Romans, nearly a millennium earlier, who were the first to take full advantage of the abundance of this readily available resource for the purposes of buildings.

The Romans, who first came to settle Britain in 43 CE, made good use of the plentiful supply of flint they found at their disposal in southern and eastern England. In the Iceni territory of what is now Norfolk locally sourced flint was combined with Roman brick in the walls of their administrative centre at Venta Icenorum a few miles south of Norwich. Flint was not only used for buildings. As compulsive road builders, Romans understood the value of the stone's hardwearing properties and made use of it for surfacing roads like the Icknield Way that ran between Norfolk and Wiltshire following the chalk – and consequently flint-rich – escarpment of the Chiltern Hills and Berkshire

Downs. They also took advantage of the stone's local abundance for widening and reinforcing existing tracks like the Peddars Way that connected the Icknield Way with their military stronghold at Brandonum (now the village of Brancaster) on the northwest Norfolk coast.

In truth, it is uncertain whether the Peddars Way existed before Romans arrived in Britain. The Icknield Way is undoubtedly of ancient origin but the route of the Peddars Way is suspiciously Roman-straight and gives more the impression of a soldiers' marching route than a solidification of a pre-existing highway. What would undoubtedly have existed would have been a selection of more meandering prehistoric ways that followed the contours of the landscape through west Norfolk to connect the Icknield Way with the North Sea as the coast swung south into The Wash.[9] These routes would likely have been previously used for trade in the Bronze Age, and perhaps even earlier in Neolithic times to take high quality flint from Grime's Graves to Norfolk's northwest coast. The Roman road builders, with their obsession for linearity, likely superimposed their roads upon existing routes when it was convenient to do so, although they were not averse to bypassing them whenever they considered it appropriate. Where the ancient ways followed the easily navigated ridges of the dry chalk it was repeated footfall that revealed the flint that lay beneath the surface. Where Romans constructed all-weather roads upon a raised earth foundation or agger, flints – or indeed any suitable and conveniently available stone – were employed to consolidate the surface.

*

Beyond East Anglia, the extensive walls of the Roman town of Calleva Atrebatum at Silchester in Hampshire are largely constructed of flint, as are the well-preserved walls of the fortress at Portchester, a third century CE 'Saxon Shore' fort on the south coast overlooking Portsmouth harbour. Another Saxon

Shore fort stands at Burgh Castle in Norfolk, a little way south of Great Yarmouth, although there is still uncertainty whether the Gariannonum of historical record was located here or on the other side of Breydon Water at Caister-on-Sea.

The fortress at Burgh Castle has impressively solid walls of layered Roman brick and flint pebble. Although now semi-ruinous, the fortress, which occupies a lofty position over the extensive reed beds of the Halvergate marshes at the confluence of the Yare and Waveney rivers, still conveys an impression of impenetrability. A mighty structure seen close up, the fortress is perhaps even more impressive when viewed from the river itself – a domineering martial presence on a bluff. Even so, like the city walls of Norwich, the fortress turned out to be a white elephant as it had been constructed against the threat of invaders that never arrived (not on Roman watch anyway). It was finally abandoned when the Romans left British shores in the fifth century.

The Roman guards may be long gone, but Burgh Castle still seems to retain a sense of their presence. The ruins have an eerie atmosphere that seems to extend across time. This ghostly ambience did not go unnoticed by aforementioned chronicler William A. Dutt, who writes:

> *One did not need to be an antiquary to appreciate a structure which had survived the passage of two thousand years, and near whose walls lay the dust of men who had walked the streets of Rome in the days of Claudius. When the shadows lurked under the bastions and the churn-owl came out of the woods to hawk for moths about the watch towers, the spell of the place grew upon me.*[10]

Some believe that the Roman fortress later became the location for Cnobheresberg, the seventh-century monastery of Saint Fursey, the first Irish missionary to southern England, although archaeological excavations carried out between 1958–61 failed to find any physical evidence of a monastic settlement. Such a lack of evidence has been explained by Louis H Dahl, who wrote a 1913 book on the subject, as simply a matter of the insubstantiality of the monastery buildings: 'It was constructed with great rapidity (velocissime), probably with no pretensions to stability and permanence, hence, the disappearance of all monastic traces.'[11] Whether or not this was ever the true location of the monastery, Christianity eventually did find a permanent foothold here at England's easternmost edge. Tucked away in a copse of trees north of the fortress, close to village of the same name, is the Church of St Peter & St Paul. This, like the fortress, is constructed largely of flint. The church is relatively modest, built mostly between the thirteenth and fifteenth centuries, although its round tower is clearly older, raised in the late eleventh century around the time of the Norman invasion.[12] The walls contain many repurposed Roman tiles taken opportunistically from the nearby fortress. The Perpendicular-style windows are largely glazed with Victorian stained glass; on the south wall is the 'Fursey window' depicting the Irish saint, which demonstrates the church's own conviction that the missionary had once preached here at the mouth of the Yare estuary. According to the church's website, 'The church has strong links with the Norwich based Ecumenical "Fursey Pilgrim" group who come on pilgrimage each October.'

Before building forts to guard their East Anglian territory at Gariannonum and Brandonium on the Norfolk coast, the Romans founded a *castrum* and military garrison on a former Iron Age defensive site at Camelodunum (Colchester) in the south. The administrative centre that grew up around this would become the first capital of the newly won Roman province of Britannia. It could also claim to be the oldest recorded town in Britain, a phrase still used on heritage road signs around the modern Essex market town that remains a garrison town to this day. In 60 CE Camelodunum, Londinium and Verulamium (St Albans) were attacked and burned in spectacularly vicious revenge attacks led by the Iceni leader Boudicca. Boudicca had been enraged by Roman duplicity on the death of her husband Prasutagus, which involved the annexation of Iceni territory, a personal flogging and the rape of her daughters. Whatever happened to Boudicca remains unclear, and the warrior leader's final resting place has never been identified, but in the century following her defeat, the Romans developed the Iceni tribal centre of Venta Icenorum (literally 'market of the Iceni') at Caistor St Edmund, a few miles south of Norwich.

Venta Icenorum, the largest *civitas* in East Anglia, has all the hallmarks of Roman culture – a grid layout with metalled streets and substantial walls of flint and brick. The town lies close to the banks of the River Tas in a prehistoric funerary landscape of Bronze Age barrows, earthworks and ring ditches. These days the surrounding fields are mostly given to growing crops or use as grazing pasture but this superimposed on an ancient landscape of chalk and flint. Several Neolithic and early Bronze Age flint working sites have been found in the vicinity,[13] and close to Venta Icenorum is a large modern-day quarry where both chalk and flint are still extracted. Out of sight, out of mind: the deeply gouged earth and dirty white scar created by the business of mining cannot be seen – or even guessed at – from the road that passes the entrance.

A couple of miles from here at Norwich's southern edgeland, close to where the River Tas has its confluence with the Yare, is the Arminghall henge. A late Neolithic henge monument consisting of a circular enclosure with two concentric rings of ditches and a horseshoe-shaped arrangement of timbers, this was only discovered in 1929 when distinctive tell-tale crop marks were observed from the air. Subsequently dated to 2450 (+/- 150) BCE, the henge was found to be orientated to the mid-winter sunset – a calendar etched into the landscape for ritual or agricultural purposes, or perhaps both.[14] An array of flint tools and blades cores were discovered on excavation of the site, although these, curiously, are of an older Mesolithic date and appear to be unrelated to the construction of the henge.

The site today is a relatively unremarkable place, little more than a low rise in a field of stoically grazing horses next to a large electricity substation. Any birdsong that might be heard is underpinned by the drone of commuter traffic washing over from Norwich's inner ring road on the other side of the river. It requires considerable imagination to squint back through the mists of past millennia and envisage how this same landscape would have looked to those who constructed the henge, and how the completed monument must have impressed those who came to attend whatever rituals were carried out here.

The timbers are long rotted away, and so is the deeper meaning of it all. There is still plenty of flint, though. Moles are habitually active here and pieces of flint can be seen glinting on the cairns of dark loam that the burrowing animals leave behind. Most, of course, are nothing but plough-shattered shards. Even so, in this ancient landscape turned modern-day edgeland, there must still be fragments of Mesolithic worked flints that, after thousands of years laying underground, wait to be held by a human hand once more.

THREE
ON THE BEACH

West Runton – Norfolk

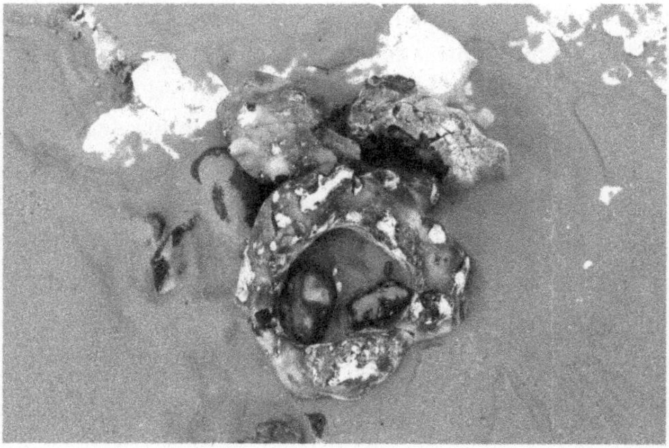

Flint – what is in a name? The English word flint derives from the Proto-Germanic '*flintaz*'. This most likely derives in turn from the Proto-Indo-European word '*splind*' – meaning to divide, split or splice. Flint is an ancient word that even in its modern pronunciation would have been understood by Anglo-Saxons more than a millennium ago, as would 'fire-stone' – *fýrstán* – the word that relates to the stone's fire-making properties. The word is identical in Old, Middle and modern English, German, Dutch and Danish. Flint – /flɪnt/ – a single syllable utterance, hard and sharp-edged in the mouth like the stone it represents. It is a name not only associated with a singular type of ultra-fine grained quartz but also to place names and surnames, which in turn might relate to a source of flint or the character of an individual.

What of flint as a place-name, a location? There is, of course, a town in North Wales that will answer to the call: Flint, in Clwyd, which even has the former county of Flintshire named after it. Flint Castle, though, constructed by Edward I as part of his 'Iron Ring' to encircle the Welsh, is not built of flint at all but of local sandstone. Across the Atlantic in Michigan, USA there is another Flint: a debt-ridden former car manufacturing city that like its Welsh namesake has no connection with the local geology, nor does the badly polluted Flint River that flows through the city and provides its dubious water supply. In contrast, the Flint Ridge Ancient Quarries and Nature Preserve further south near Brownsville, Ohio is precisely what it claims to be: a memorial park to the many pre-Colombian Native American flint quarries that once stretched for nine miles along the ridge here. The flint extracted here was a highly valued resource that was traded as far west as the Rocky Mountains and south to the Gulf of Mexico.

Flint can be used as a simile to describe human traits and temperament. The Bible (*Isaiah* 50:7) has Jesus pronounce: 'Therefore have I set my face like flint, and I know I will not be put to shame,' using the term to describe strength and determination. It can also be used to mean hard and inflexible or of a severe unyielding character, especially in regard to 'flinty' eyes or 'flint-like' hearts. In all cases, the reference to flint is to make as a comparison of the stone's physical properties with that of human character. Generally speaking, it is not used flatteringly.

More prosaically, the term flinty is used as an adjective to describe the character and flavour of wine. To describe a wine as flinty is to suggest a taste that is crisp and sharp, dry and perhaps a little austere. It implies a degree of what wine aficionados refer to as minerality, a connection between flavour and the geology of the *terroir* where the grape is grown. It is often associated with the crisp white wines of the Loire Valley, the much admired Sancerres and Pouilly-Fumés of this flinty-soiled region. It also hints at the smoky smell of gunflint against steel. Unlike its use in describing human character, to suggest that a wine is flinty is unquestionably a compliment to the winemaker. It is no coincidence that one of Norfolk's most recently established vineyards has chosen the name *Flint* for its business. In the world of viniculture it is a name that suggests finesse and class. Flint Vineyard, naturally enough, is located on the flint-rich soil of the Waveney Valley on the Norfolk-Suffolk border.

*

In geology, flint is a curiosity, a unique type of stone that is born within another. It forms as nodules or layers of hard stone within chalk, a much softer rock. Like chalk, it is produced from the residue of living creatures from a former geological age. What was once organic slowly becomes mineral. Chalk's formation from the layered accretion of the shell of sea creatures is easy enough to envisage. Flint, though, requires more tangential imagining.

For many years the exact mechanics of its formation was a matter of debate in geological circles. I remember, back in my university days, an Environmental Science field trip to the Norfolk coast. We were looking at rocks on the beach and someone asked the lecturer about the line of flint nodules that were visible in the chalk cliffs above us. 'How is it formed? Well, the jury's still out on that one.' He went on to describe how dissolved silica from Cretaceous-era sponges formed layers in the chalk but admitted that no one could really agree what the exact process was. I gave it little thought at the time but now I wonder: has our geological knowledge progressed since then? Can we now be more certain about precisely how flint is formed?

Flint: its Origin, Properties & Uses (1972), by Walter Shepherd, considers in great detail the possibilities of its formation. A lengthy chapter on 'The Segregation of Flint' painstakingly considers 'syngenetic', 'penecontemporaneous' and 'epigenetic' theories, in addition to what Shepherd calls physical and chemical theories, over its forty pages.[1] The theories outlined here can be divided broadly into three groups according to the actual time frame of the transformation of dissolved silica into flint. For most readers, including me, this is overcomplicating things considerably. Put in the simplest terms, it is now generally thought that flint nodules result from dissolved silica released from the exoskeletons of microscopic sponges accreting and crystallising in the void left by burrowing crustaceans. It is the shape of such burrows that give flint nodules their curiously knobbly and sometimes perforated organic form, a metamorphosis from silica sludge to sinuous, amorphous lumps of very hard rock.[2] In some instances, the same dissolved silica flowed into the joints and bedding planes of the chalk bedrock to solidify, or into the spaces occupied by dead sponges and sea urchins, the liquid silica absorbing their skeletons to trace the outline of their shape on solidification. This, currently at least, is the established scientific reasoning – a best-fit explanation of the processes involved. Personally, I much prefer the beautifully succinct explanation given by Alan Garner:

> *The weight of half a mile of chalk pressed the silica of the shells and scales. Larger creatures burrowed through the ooze, and the silica followed, filling the tunnels and burrows; settled, and was hard, a cast of where they'd been.*[3]

The beauty that is flint is not a stone that gives up its secrets easily. Did those roaming Palaeolithic tool makers even wonder about such things? It seems unlikely, although no doubt they would have respected the stone's utility. The sweet taste of honey, the thrill of a companion's skin and the regular rotation of heavenly bodies in the night sky were probably far greater marvels, although such things can never be taken for granted.

*

Perhaps, to put things in perspective, it is best to go back to the very beginning. Around 14 billion years ago there was an explosion, an unimaginably big one. What sparked this, the 'Big Bang' – the first event of every occurrence that would ever take place in time – is unknown, and perhaps by its very nature, unknowable. Some might say it began with a word; or rather, *the* word – the divine breath of the logos. What is certain is that it wasn't caused by a spark made by striking a flint, because this stone, along with solid matter of any kind, was yet to come into existence... as, of course, was any consciousness to experience it or any sort of language to describe it. Following the explosion everything was pure chaos for a while, especially for the first twenty minutes or so when nuclear fusion created and destroyed particles of matter in the passing of the most infinitesimal fractions of a microsecond.

Between 200–300 million years later – a year being the measurement of the time it would take an obscure, yet-to-be created planet to rotate around an obscure future star in an obscure as yet uncreated galaxy – the first stars started to shine as they transformed hydrogen to helium at their core. I am being flippant here but how else might I conflate the infinitesimally brief and indescribably long? How can we measure such hugely variable passages of time on the same linear scale? But even time is not quite as reliable a dimension as it might seem; nor as linear. Gravity is capable of bending light, it can also curve space-time. Time runs slowest where gravity is strongest.

Around 4.6 billion years ago, the solar system that would contain our tiny speck of a planet came into existence. The Earth was then a spinning lifeless sphere of rock full of fire and foul gases. Some of these gases would eventually condense to leave behind water and a potentially life-supporting atmosphere. A billion years later, the first carbon-based life forms would develop in the seas that then covered the planet's surface until one day, holding breath and metaphorical noses, some of the more plucky life forms were brave enough to emerge from the water and try their luck on dry land.

Over half a billion years ago, a massive diversification took place, which created most of the major animal and plant phyla that are still in existence today. Another two hundred million years later, giant reptilian creatures would roam the dry land, many of which, their true appearance surmised only from extrapolation of their fossilised bones, would delight children of the future with their monstrous forms, imagined antics and the funny names we have given them. Eventually, 66 million years ago, these dinosaurs, along with three-quarters of all living species, perished in a calamitous event precipitated by a chance meteor strike just offshore on what is now Mexico's Yucatan peninsula. The event was a singularity that fatally changed the conditions required for life, the asteroid's impact forcing enormous quantities of sulphur and soot into the atmosphere to black out the sky and cause freezing temperatures that persisted for at least a decade.[4] Mercifully, those dinosaurs that could take to the air and which would later evolve into birds were spared. This was the so-called Cretaceous-Paleogene (K-Pg) extinction event, the cataclysmic upheaval that ended the Age of Reptiles and heralded the Age of Mammals. The Mesozoic era was over, dead and buried. Long live the Cenozoic.

Meanwhile, in the Earth's crust beneath both land and sea, layers of rocks were slowly but inexorably shifting as others were still being formed. Whole continents were on the move, breaking away from the single supercontinent that we have retrospectively dubbed Gondwanaland. In the late Cretaceous period, North America, Greenland and Eurasia were still mutually attached. South America and Africa had only recently gone their separate ways and Australasia was still connected to Antarctica. The entire Indian subcontinent was a free agent, a solitary triangular island adrift in the ocean that Western cartographers would eventually name after it.[5] As the atmosphere warmed, and the shallower seas started to dry up, the tiny skeletons of plankton that lived in them died and fell to the bottom to build up into a thick white ooze of calcareous material. Over millions of years trillions of these microscopic skeletons accumulated into thick layers of calcium-rich deposits, which hardened into the white rock that we call chalk. In many instances, when the chalk underwent heat and pressure, this same rock metamorphosed into the harder rock we know as limestone.

Deep down in the chalk, like shiny beads of buried treasure, lay bands of another rock formed from the silica squeezed out of tiny sponges, rather than from the calcium of seashells. This was flint. By the time dinosaurs were wiped off the face of the Earth to leave nothing but rare giant footprints and intriguing, often faith-challenging fossilised bones, flint was firmly lodged in the chalk strata. Over time, a portion of this flint would be shifted by tectonic movement and flowing water and ice to be widely scattered across the land's surface as well as on shorelines where a twice-daily rattling by the tide would chime its melody to the rhythm of the moon. In time – in a shallower layer of deep time that lies closer to the surface – another sort of music would be made. This would be the sound created by those early humans who picked up the stone and struck it to make tools and fire. A human rhythm and a percussive, sharp-edged note – *flint* – all attack and no decay, its name almost onomatopoeic.

*

After oxygen, silicon is the second most common element on Earth. Flint is composed of silicon in the form of silicon dioxide and combines the two commonest elements in chemical union. While oxygen is the most important of elements for obvious reasons, silicon, despite its modern use in gel and electronic wiring, is ironically one of the least used (the iron in ironically is, as a matter of interest, the fourth most common element and one that has proved itself pretty useful over the last two or three millennia). Flint's silica is biogenic in the sense that it originates from living organisms. Like the chalk that envelops it in its natural state, flint has made the transition from organic to inorganic, from living thing to mineral.

In geological parlance, flint is a sedimentary cryptocrystalline form of quartz, a variety of chert that occurs in chalk or certain types of limestone.[6] While quartz itself is silica (silicon dioxide – SiO_2) in tetrahedral mineral form, chert is much closer in nature to flint in being extremely fine-grained and of organic origin. Flint and chert are sometimes confused or, as is the case with some North American geologists, the terms are used interchangeably, although in Britain the word flint is used to describe the widely found variety of chert that is found in Cretaceous Chalk. Chert tends to be paler than flint in appearance on the whole, a little softer and lighter and less dense. Because of their very fine grained nature both chert and flint break to produce a conchoidal fracture that can be used to make sharp-edged tools. Naturally occurring deposits of chert were often used in the past for tool-making in areas like southwest England where deposits of flint are scarce.[7] It is flint though, quite literally, that has the edge.

To fracture conchoidally means to break with a smooth curved surface without following any natural planes of separation. It is the same manner in which glass breaks: a particularity of material, like flint or volcanic rock such as obsidian, which has no obvious crystalline structure. Obsidian shares a number of similarities with flint and might be considered as a sort of geological cousin, although its origin is very different as it is formed by volcanic activity. Obsidian is produced from lava that as well as a high silicon content is rich in elements like oxygen, potassium and sodium. Like some varieties of flint, obsidian is black in colour; like flint, it is virtually grainless and fractures to leave a razor sharp edge. In some pre-metal working cultures where flint was scarce or absent, it was obsidian that took the place of flint in manufacturing blades and weapons. Because of its origin in lava flows obsidian tends to be found close to volcanoes, although is relatively uncommon in Europe. Large deposits can be found in central Anatolia in what is now Turkey, a source that was used widely throughout the prehistoric Near East. In the volcano-rich New World, obsidian was a valued resource in several Pre-Colombian Mesoamerican cultures. Mayan, Toltec, Olmec and Aztec civilisations all made great use of obsidian's inherent razor sharpness for use as tools for butchery and also for some quite terrifyingly brutal weaponry. Obsidian also had its non-utilitarian uses as jewellery and grave goods as well as providing tools for auto-sacrificial bloodletting. Its smooth reflective surface also meant that obsidian could serve as a mirror.

Mirrors were an important aspect of divination in Pre-Colombian cultures. These were often liquid, as in a pool of water, or a reflective surface like polished iron ore but by the time of the Postclassical period that preceded the Spanish invasion, obsidian was often the favoured means of producing mirror-like objects. Aztec rulers oversaw their subjects by means of a double-sided obsidian mirror; the belief being that one side would enable the ruler to see how his subjects were behaving, while the subjects were able to see themselves reflected back on the opposite face.[8] Such mirrors sometimes travelled far. Dr John Dee, court astronomer to Queen Elizabeth I, possessed an obsidian scrying mirror and recent geochemical analysis suggests that it had its origin is in the Pachuca region of Mexico, a major source of obsidian during the time of the Aztec Empire.[9] Little is known of how Doctor Dee obtained the mirror but the object was most likely brought to Europe as booty by Spanish conquistadors in the 1520s. Rather than using it for keeping an eye on his queen's subjects, Dee, an enthusiastic magician as well as astronomer and mathematician, adopted the mirror as a 'shew stone' for his own personal occult research. After Dee's death the mirror eventually passed into the possession of antiquarian and politician Horace Walpole. It now resides in the collection of the British Museum.

There is precious little obsidian in Britain but an abundance of flint, especially in the south and east of England where the underlying rock is chalk. Much of the flint lies deeply buried underground but there is plenty on the surface too, littering the ground on ploughed farmland or as pebbles or shingle on the beach.

Scattered along the East Anglian coast, separating idyllic expanses of golden sand at places like Holkham, Great Yarmouth and Southwold, are melancholic stretches of flint shingle. These places hold an atmosphere more given to philosophical pondering than the family-friendly sandy beaches where buckets, spades and swimsuits prevail. The pristine golden beach seems to have a strong hold on the human psyche, offering notions of sybaritic life and earthly paradise. Some of this, I suspect, is down to marketing – the sun-kissed strand of the glossy travel brochure. Part of it is also because of what beaches signify as symbols of carefree holiday simplicity – a break from the drudgery of everyday life. Shingle beaches tend to engender a different mood, and whatever the weather they don't offer this kind of beach holiday experience. Their appeal is more for fishermen, birdwatchers and solitary walkers, although on beaches of this sort walking itself is more of a challenge. As Jules Pretty observes of shingle walking in his almanac *The East Country:* 'It was a six-mile walk to advance two. The Suffolk shingle mile has always been a long one.'[10] And so has the Norfolk shingle mile, and the Kent and Sussex shingle mile too – the rate of exchange for stone-crunching effort is much the same whatever the location.

I would walk many shingle miles researching for this book, returning to beaches I already knew in Norfolk and Suffolk as well as venturing to others, hitherto unknown to me, which lay further afield. But first I wanted to return to somewhere I had been many times before. A place where the chalk meets both shore and tide, where flint could be seen in all its knobbly glory lined up in the cliffs and scattered across the beach. A place of eccentric geology and extraordinary forms, of fossils and soft crumbling cliffs: the stretch of shoreline that lies between Cromer and Sheringham on the north Norfolk coast.

*

West Runton is probably best known for two things: the West Runton Pavilion, a fabled one-time music venue that a number of well-known rock bands used to favour for a warm-up gig at the start of a British tour, and the so-called 'West Runton Elephant'. While the West Runton Pavilion hosted a broad spectrum of late 1970s acts that include The Cure, The Clash and Iron Maiden, as well as an infamous Christmas Eve performance by The Sex Pistols, the story of the West Runton Elephant is more recent but will likely go on to have greater longevity.

The 'elephant' was discovered at the base of the cliffs in West Runton in December 1990 after a storm had exposed one of its fossilised bones in the cliff face. The bone was identified as the pelvic bone of a steppe mammoth (*Mammothus trogontherii*), a creature that lived in the region between 600,000–700,000 years ago. A year later, following another winter storm, more bones were found at the same location. A large scale excavation took place in 1995 once further funding had been organised. What was eventually unearthed was the most complete fossil steppe mammoth to be found anywhere.[11] The mammoth was a member of the largest elephant species ever to have lived, and with the exception of dinosaurs, the largest animal ever to have lived on land.

When the mammoth walked this coastline Britain was still attached to the Eurasian continent by the expanse of low-lying and now-submerged land known as Doggerland. The animal's contemporaries were spotted hyena, giant beaver and extinct species of rhino and giant deer. Given recent archaeological evidence, the same territory may also have been shared by some of our distant ancestors, as at the same time this same stretch of coastline was also roamed at the same time by small bands of Homo antecessor or 'Pioneer Man', who according to fossil footprint evidence, were living further along the coast at Happisburgh.

Although it is best not to hold out too much hope for finding anything as grand as a fossilised mammoth, West Runton has more reliable treasures. When the tide is at its lowest, it is easy to spot paramoudras perched upon the exposed chalk on the beach. Resembling weirdly shaped, bucket-sized doughnuts – or even sometimes lavatory seats – paramoudras are flint trace fossils that have built up around an unknown and long-vanished worm-like creature, which for the sake of paleontological record is referred to as *Bathychnis paramoudrae*.[12] The only memorial to this creature's onetime existence is the fossilised space it leaves to posterity.

I went for a walk on West Runton beach with a couple of friends who lived locally. It was a cold late winter morning, and even colder on the beach where an eye-watering north-easterly was coming straight off the sea. The north Norfolk coast can sometimes feel as if it is at the edge of things – a running out of land. And it is: while only at about the same latitude as Berlin, head due north from here and there is no more solid land until reaching the Arctic pack ice. On a chilly winter's day it can almost seem as if the rafts of wind turbines that lie offshore are blowing the wind back at you like cooling fans.

The tide was completely out and yet to turn; the lower beach, a simple palette of flat wintry tones: the milky white of the exposed chalk in contrast with the darker flints that rested upon it, their shiny black cores coated by a bone-coloured cortex that in some cases was stained with rust-red streaks. The uneven chalk pavement at our feet was interspersed with small rock pools and shallow channels of water. Many of the larger flints that lay scattered were paramoudra – tubular in shape and either hollow in the middle or filled with chalk like a sculptured vol-au-vent. Given their appreciable size and weight, I wondered how many were regularly pilfered from the beach to take pride of place in a garden rockery or for use as a doorstop.

The name paramoudra is Irish, deriving from the Gaelic *peura muireach*, meaning 'sea pears'. They have also been called 'ugly Paddies' in the past, which seems a little harsh, even racist. They are beautiful in their own way.[13] Their Norfolk name of 'potstones' makes more sense, as some of the better formed ones could easily be adapted to serve as plant containers. Paramoudra, like all flints, are actually pseudofossils. They are generally thought to be fossilised barrel sponges but the precise process of their formation is not fully understood. Geologists have separated paramoudra into a number of categories but essentially there are two main types – *paramoudra typica* and *paramoudra magna*. The former are knobby hollow barrel-shaped flints up to about 60 centimetres wide, while *paramoudra magna*, as their name suggests, are larger and consist of a circle of flints up to a couple of metres in diameter.

We moved closer to the cliffs to look for fossils among the pebbles. Christine, who comes here almost daily, already had quite a collection of belemnites and it was only a matter of minutes before she found a few more. Once I got my eye in, I managed to identity a couple for myself. Like little stone bullets that taper elegantly at one end, the fossils stood out from the surrounding pebbles with their smooth elongated shape. The fossilised internal shell of an extinct but once common cephalopod, they are the stone-cast ghosts of squid that once swam in the warm seas that prevailed in the Jurassic and Cretaceous periods.

Pleased with my find, I carried on looking a while longer. I did not find any more belemnites but just as I was about to give up and catch up with the others I saw something else among the shells and shingle. An echinoid, a fossil I was already familiar with as I already had a few lined up on a shelf at home.

Echinoids are fossilised sea urchins that are commonly associated with Jurassic and Cretaceous deposits, especially Late Cretaceous chalk. They are unmistakable because of their distinctive lampshade shape and five-fold symmetry. The original spines of the urchin are invariably missing but no matter how damaged or eroded by tidal action the pentagonal symmetry of the shell is usually plain to see. The one I had found was broken and had a sizeable chunk missing but even so it was an intriguing object – a tangible connection with the deep past that sat satisfyingly heavy in the hand.

The thing about an echinoid is that it is a form that defines an absence. Effectively it is a flint cast – a faithful reproduction in silica of the inner space of a 100 million-year-old living thing. The flint itself is a pseudomorph, a secondary or 'false-formed' rock deposited in the place of another rock (chalk) that has been removed. An echinoid is, as writer/stonemason Alex Woodcock observes, 'a twice-removed thing: like making a mould in which to catch something (say, plaster), casting it and then slowly replacing the plaster cast with another material.'[14] The resulting fossil represents an extraordinary transformation of matter over tens of millions of years. From white chalk to black silica, calcium carbonate to silicon dioxide: an echinoid is chemical alchemy frozen in stone.

The echinoids I had at home had been given to me by a neighbour more than thirty years earlier. The neighbour, a friendly man called Russell, who used to work at a chalk quarry close to Norwich, came across objects like this almost daily. Russell, who often stopped to chat, had me marked as the sort of person who might be interested in such things and one day he presented me with a handful of them over the garden wall. I was happy to receive them but did not give them too much thought back then. Even so, I liked to pick them up from time to time and the more I became interested in the weird and wonderful world of flint, the more I found myself wanting to handle them. There is something tactile about a fossil of this sort that is satisfying yet difficult to identify. Is it the purity of its curved form, the five-fold symmetry, its agreeable compactness and weight? To hold one in the hand is to connect with the ghost of something that existed long before our own puny historical span, something that comes from beyond the veil of everyday experience.

Echinoid fossils have been collected for millennia and have been found at archaeological sites that range in age from the Palaeolithic to as recent as Anglo-Saxon. They hold a place in resurrection myths in both Celtic and Norse mythologies and have acquired folk names, like shepherd's crowns, fairy loaves and thunderstones, that give an indication of the sort of folklore that is associated with them. They have also been called lucky stones, policeman's helmets, knee caps and even bishop's knees. Echinoids have also often been found as part of grave goods and appear to have long held some sort of significance as a connector between the dead and the living. The same fossils have even been found embedded around the windows of a church in the estate village of Linkenholt in Hampshire's North Wessex Downs, a nod perhaps to a pagan past in the same spirit as those representations of the Green Man that are often seen in medieval churches. The church in question, St Peter's, is not so old – mid-Victorian, in fact – and so it might be assumed that the echinoids had been carefully removed from the original medieval church in order to be reinstated in its Victorian successor. The fact that a similar arrangement of fossils can also be seen around a window of the church's contemporary, the one-room schoolhouse next door, however, puts this in doubt.[15]

A particularly spectacular example of an embedded echinoid comes from a flint hand axe found at an Early Palaeolithic site at Swanscombe in Kent. Here a whole echinoid has been incorporated into the face of the flint tool, covering about one-third of the surface. To include the fossil in this way was clearly a deliberate choice, although whether this was merely a convenient aesthetic embellishment by the *Homo heidelbergensis* maker, or if the fossil was believed to confer some sort of special power, we will never know.

Fossil sea urchins have been found at burial sites throughout the archaeological record. An excavation of a Bronze Age grave on Dunston Downs in Bedfordshire has revealed the eerie spectacle of a ring of around two hundred echinoids surrounding the skeletons of a young woman and child. We can presume the circle was put in place to provide some sort of spiritual protection in the afterlife but we cannot be certain. The practice of burying the dead with echinoids appears to have continued well into Anglo-Saxon times. At a cemetery at Westgarth Gardens in Bury St Edmunds, Suffolk, an Anglo-Saxon grave was discovered in which, in addition to brass brooches on her shoulder and glass beads around her neck and waist, a woman had a fossil Echinorys, a type of echinoid, clutched tightly in her right hand.[16]

More generally, the fossils have been associated with good luck. They have also served as amulets – some Iron Age echinoids have been found with metal clasps attached – and because of their easy portability and perceived protective power were often carried on journeys to afford the bearer safe passage.[17] In more recent times, echinoids have been linked with quite specific attributes. In Suffolk, fossil echinoids, which were known locally as fairy loaves because of their supposed loaf-like shape, were placed beside the hearth in the belief that homes with these in place would never go short of bread.[18] In Sussex, they were traditionally positioned on window ledges to bring good fortune to the household, and on the North Downs their presence in a pantry or dairy was believed to prevent milk from turning sour.[19] Echinoid fossils were also thought to protect homes against flooding and lightning strikes. South Downs shepherds used to carry them for this same purpose as a lightning strike during an upland storm was always a tangible threat.

Myth and legend intertwine. Fresh myth wraps itself around pre-existing fable. In Denmark, where they were also associated with thunderstorms, echinoids were thought to be forged at those places where Thor's bolts of lightning struck the ground.[20] The same meaning was attributed to flint axes found in the Danish countryside, while in the southern islands it was a different fossil, belemnites, that were considered to be 'thunderstones'. As the palaeontologist and fossil expert Dr Ken McNamara asserts: 'In Europe fossil sea urchins and belemnites conveniently became bit players in powerful Norse mythologies… as they brought with them powers bestowed by the dragons or by the gods.'[21]

What was it in particular about these fossils that gave them such magical meaning? Their five-fold symmetry is certainly beguiling. It is a symmetry that is common enough in nature, in the arrangement of flower petals, in fruits – the cross-sections of apples and okra – in starfish, even in viruses. Five-pointed symmetry has been associated with magic and mysticism since ancient times. This same symmetry defines the shape of the pentagon, the magical symbol upon which Leonardo da Vinci superimposed the spread-eagled human figure. Spread out our arms and legs and we all become pentagons. Our fingers and toes are already arranged in convenient sets of five. A five-pointed star, the star of communist revolution, can be drawn with just five pencil strokes and a steady hand. Of the Greeks' five Platonic solids – earth, air, fire and water – pure celestial matter was the fifth. In the epic Middle English poem *Sir Gawain and the Green Knight*, five is associated with spiritual purity and virtue. The hero goes into the world 'fully faithful in the five ways five times over'. The shield and sword he carries is adorned with the 'symbol of Solomon' in the form of a pentacle or what is referred to here as 'the endless knot'. The knight is 'flawless in his five senses', his five fingers 'never at fault', the five wounds of Christ and 'the five joys of Mary', his faith.[22]

The same pattern of five lurks in architecture and tile making, in Christian, Islamic and Ancient Egyptian iconography.[23] Sir Thomas Browne, the seventeenth-century Norwich physician and proto-scientist, made much of the five-fold symmetry he observed in nature, in particular the five-sided lozenge shape he called the quincunx, a shape that combined to form a pattern of infinite repetition and which occurred widely in nature, especially in plants, and in objects of human invention like medieval brickwork.[24] Five-fold symmetry exists in 19th-century military fortifications too and, of course, Pentagon is both the name of the Virginia headquarters of the United States Department of Defense and the shape of its architectural form.

Beyond the symmetry, such fossils were probably revered simply because of their unworldly strangeness. A once-living creature whose form has been frozen in stone, a dead thing that in its deathly beauty reveals a clear and distinct pattern, the world of such mysterious objects was antediluvian, deeply mysterious and unknowable. A Neolithic flint miner, perilously digging down into the chalk, would have been stopped in his tracks by their elegantly shaped, other-worldly splendour. It is little wonder then that their magical appeal continued long beyond the age of stone.

What pleased me about the broken echinoid I found on the beach at West Runton was that it had a direct geological connection to the ones I already possessed. West Runton beach lies at the northern end of the Beeston Chalk, a long belt of chalk that runs from the Norfolk coast to just south of Norwich. The fossils given to me by my former neighbour came from a quarry at Caistor St Edmund – a quarry that extracted stone from the same belt of chalk that could be seen exposed at low tide at West Runton.

We continued our walk along the beach as far as East Runton. Silhouetted in the distance we could make out the pier at Cromer and the tall tower of the town's parish church. East Runton, which was once a fishing village but had since become a spill-over from its resort neighbour, was now mostly given over to the caravans and campsites of the holiday industry. The cliffs were striking: white chalk banded with folded glacial till, crag, clay and conglomerates – what is referred to as Contorted Drift. But geology had been turned upside down here: the past was out of sequence and in places the younger beds of the Pleistocene lay *below* the much older Cretaceous chalk, a phenomenon illustrating how easily natural sequences can become convoluted.[25] In other places where the chalk came down to beach the rock was studded with neat bands of flint, the black nodules standing out like currants in a cake, as clear as anything you might see in a text book. It was the sort of feature that an earnest teacher might point out to students to remind them of something they had been taught in a geography lesson.

FOUR
FLINT MUSIC

Blakeney Point – Norfolk | Orford Ness, Shingle Street – Suffolk
Dungeness – Kent

Listen! You hear the grating roar
Of pebbles which the waves draw back, and fling,
At their return, up the high strand

Matthew Arnold *Dover Beach*

Flint is more widespread than you might imagine. Many beaches around the British coast, particularly those in the south and east of England, are composed of pebbles derived from the same hard sedimentary rock that lies layered in the chalk. Flint is an apple that sometimes falls far from the tree, and just because there are no obvious sources nearby does not mean that it cannot turn up among, or even dominate, the multiplicity of stones found along a shoreline. Although glaciers are no longer active in temperate climes, other forces are at work rearranging the landscape and shifting the rocks around. On the coast, it is the sea that is the highway by which flint is transported from its origins in the chalk; longshore drift is the engine that drives it.

Longshore drift is the geomorphic process by which material is transported along a coastline parallel to the shore. The material is invariably sediment of some sort – anything from minute particles of silt and sand to large boulders. Longshore drift is the engine behind many of the eccentricities of much of the British coastline, especially in the south and east of England. It is both the gift that gives and the scourge that takes away. Coupled with worsening climate change, it is the reason why large tracts of the East Anglia coast have been dramatically changed over the years, why river ports have silted up, why cliffs and even whole villages have been swallowed up by the North Sea. It is also the process by which those large, ever-enlarging spits of shingle and sand that distinguish parts of the Norfolk and Suffolk coastline have been created.

Matthew Arnold's poem *Dover Beach*, which evokes the sound of grinding pebbles, was written in 1867 as an elegy to what he believed was the loss of religious faith and how this was losing ground to the sciences and the new evolutionary theories that had recently been championed by Charles Darwin: too much geology, not enough god-fearing, as he saw it. Arnold's poem expresses an existential fear that without sufficient faith the tide might not return. The sea of faith is seen as 'retreating to the breath/ of the night-wind, down the vast edges drear/and naked shingles of the world.'

Spits occur where material like sand and pebbles is deposited parallel to the shoreline. A typical spit broadens at its head and curls inwards towards the land where the effect of longshore drift is less impactful: a feathery growth of new land whose development observed over time can appear organic and abstractly beautiful. For a spit to form, longshore drift needs to pass a coastal inlet of some sort – an estuary or river mouth. When the conditions are right, and when the direction of drift and shoreline veer at a particular angle and factors like wave direction and currents are also optimal, spits will build over time.

Great Yarmouth, Norfolk's largest port and latter day holiday resort, sits on a spit, squeezed into the confines of relatively recently created territory that lies between the North Sea and the mouth of the River Yare. The spit that the town occupies runs north to south tracing the direction of drift along the east Norfolk coast. On the north Norfolk coast, things are different and longshore drift runs east to west. This explains the formation of the two large spits that have developed along this coastline – Blakeney Point and Scolt Head – although the latter has since separated from the mainland to become an island.

Blakeney Point extends for four miles along the Norfolk coast parallel to the coastal villages of Cley next the Sea, Blakeney and Morston. Unlike Great Yarmouth, it is mostly seals and seabirds that have sovereignty here. Composed entirely of shingle and sand dunes, Blakeney Point is a lonely yet melancholically beautiful place.

One phenomenon of longshore drift is that the further you go in the direction of the drift – west in the case of Blakeney Point – the smaller the pebbles become. This is a result of lighter sediment travelling further, heavier particles less so. The lightest particles of all, grains of sand in their billions, are blown by the wind and stacked up in dunes that are held in place by the long-rooted marram grass that grows upon them. Although the dunes are well established, by their very nature they are ephemeral and it is only Blakeney Point's manmade features that have any real sense of permanence about them, although even these appear vulnerable. In the flat watery landscape of the north Norfolk littoral, these rare landmarks can be seen from miles away. At the end of the Point stands the distinctive blue structure of the warden's house, a curved roof building that once served as a lifeboat station. Halfway along, between the warden's house and the beach car park at Cley Eye, is a small brick building known as Halfway House, a property occasionally used for residential field trips. There is no need for a more poetic description for a building on of a four-mile stretch of land that has only two dwellings to its name. Besides, no postman could ever be persuaded to deliver here.

The soundscape of the Point is a melange of bird, wind, water and stone. More specifically, it is the harsh cry of gulls and the wistful fluting of redshank, oystercatcher and curlew against a background of wind combing through sharp tangles of marram grass. Beneath this is another quieter, yet more persistent, set of sounds: the sough of waves rolling in on the beach and the percussive clink of abrading pebbles as the water rises and falls – flint music in its purest form.

Because of its unique topography and abundance of wildlife the whole territory of Blakeney Point is a designated National Nature Reserve as well as a National Landscape (formerly known as an Area of Outstanding Natural Beauty, or AONB) and a Site of Special Scientific Interest (SSSI). Other than by boat, the only access is from the beach car park at Cley Eye from where it is four calf-aching, shingle-crunching miles to reach the end of the point. Thanks to its position and wide range of habitat – shingle bank, mudflats, marsh, beach, grazing meadows – this part of the Norfolk coast has long been a bird Mecca. Few places in the United Kingdom have delivered quite so many rare species over the years; there again, few are watched over quite so diligently by birders. With expensive German optics searching for elusive Western Palaearctic migrants, it is a surveillance zone of sorts, with avian rather than human activity the prime interest.

When I was a young man I once came here to camp for the night – obviously not something that the National Trust would approve of these days. I hitch-hiked to Cley next the Sea from Norwich, trudged out along the shingle and set my small tent up close to Halfway House. It was late summer as I recall but by the time dusk came the car park had emptied and there was absolutely no one else about. I sat watching the sea allowing the solitude of the place to wash over me as I ate the sandwiches I had packed. When it got dark I climbed into my sleeping bag and tried to sleep. I dozed for a while but later woke up feeling disorientated and ill at ease when I experienced a vague sensation of sinking – or perhaps, it was drowning. It felt almost as if I were floating on the sea, buoyed only by the flinty raft of pebbles beneath me. This was, of course, partly down to the peculiarity of the location, except I was not floating at all but on solid ground, albeit at the flimsier end of the terra firma spectrum. What disturbed me most was the all-enveloping sound of the sea and the grinding pebbles. Pitched on a narrow pebble spit between two bodies of salt water, I could hear the action of the tide on both sides of where I lay. The groan of the pebbles and the swash of the tide arrived at my ears stereophonically or, as it seemed at the time, with something akin to cinematic surround sound. As time passed, both sides seemed to grow louder, as if the tide was encroaching and the water would soon wash over me. While I became further engulfed by the sound of tide and stone, unseen gulls added fly-by bird scream to enrich the mix. It seemed as if there were muffled human voices too. In my imagination they sounded wraith-like and anguished, like cries of the drowned. In the disorientating darkness, nature had grown intimidating rather than welcoming. Like a vagrant wader or a glacial erratic, I was an alien presence and I had a tangible sense of being in a place where I really ought not to be. The experience was certainly memorable. It was not until years later that I learned that Halfway House had provided inspiration for the haunted house of E. F. Benson's 1925 Norfolk-set ghost story *A Tale of an Empty House.*[1]

*

I visited Blakeney Point in early summer with Jackie and two friends who live nearby. Klausbernd is German but has lived in Cley next the Sea for more than three decades, well before the village transformed into the wealthy second-home enclave it has since become; long enough to see the spit at Blakeney Point subtly shape-shift with tens of thousands of pebble-rolling tides. According to Klausbernd, Cley is these days full of retired Cambridge professors and has more Nobel Prize laureates than you can shake a walking pole at. Neighbouring Blakeney, arguably even more gentrified, is replete with sea-loving, second-home-owning accountants. In his view, both villages are occupied by affluent folk but it is Cley that has the culture. The 'property porn' houses and cottages of both villages are built almost entirely of flint pebble; even new developments in the case of Cley, where strict planning laws are enforced to ensure architectural integrity and continuity. Flint's legacy here means expensive real estate. Klausbernd's partner Hanne is Norwegian and has only lived here permanently since she took retirement a couple of years ago. Both seem thoroughly at home in the village and have taken on some of the healthy patina that living on the coast brings for some – an unhurried, windswept outlook on life that blends well with the curious, non-judgemental view that incomers born on different shores tend to bring with them.

It had been sunny in Norwich when we left but it was slightly hazy at the coast, a day caught on the cusp as an unusually cold spring stumbled into an as yet unknown summer. Leaving the beach car park we headed west along the Point past a few beached lobster boats, scrunching through the shingle to reach a path that meandered through a tangle of glaucous shrubs at the edge of a salt marsh. Out of sight, beyond the shingle ridge to our right, we could hear the constant mineral grumble of flint pebbles grinding on the tide. Ahead in the distance Halfway House cut a lonely figure in the landscape. Beyond, in the salty haze of the far end of the Point, we could just about make out the bright blue of the visitor centre.

We were here, at Hanne's invitation, for the little terns. Hanne is one of several local volunteers responsible for keeping an eye on the birds. The terns nest at awkward and not particularly sensible places in the shingle, leaving their eggs vulnerable to high tides and careless human feet as well as prone to attack by predators like herring gulls and kestrels. A cordoned-off area of shingle enclosed some of the tern's nests, although many of the birds had already finished breeding and had moved further west to the end of the Point. The same enclosure also held a few oystercatchers and avocets, although each species seemed to be content to mind its own business. Anchored in the firmer shingle were salt-tolerant plants like sea beet, sea campion and biting stonecrop, while at the looser-stoned apex of the ridge leathery stands of seakale were in full bloom. Elsewhere, clumps of yellow horned poppy, another shingle specialist, were starting to throw up flower heads. A place that you instinctively feel should be barren, it seemed remarkable that anything could grow here nurtured by little else but sand and saltwater.

Hanne took us for a walk up towards Halfway House. The bushes were alive with restless flittering birds – a mixture of meadow pipits, linnets and reed buntings, although we knew that almost anything could turn up here at times of migration. A skylark sang somewhere overhead, little more than a high fidgeting dot to the naked eye. In the distance, across the marshes close to Blakeney Channel, we could make out the unmistakable form of a marsh harrier quartering the reed beds.

The most impressive sight by far, though, was a meeting with a pair of brown hares. One immediately made a run for it disappearing into the thigh-high Suaeda (sea-blite). The other remained, frozen in its tracks, hunched with ears flattened to its head in an attempt to make it appear smaller. More resembling a small deer than a large rabbit, with improbably long ears and soft, intelligent eyes, it was easy to understand how hares have always been revered in British and European folklore. Long gifted magical properties by those whose livelihood entails a close relationship to the soil, hares engender a strong sense of otherness. A sacred animal, a spirit familiar, a symbol of fecundity, sex and madness; a means of divination too: the Iceni warrior queen Boudicca is said to have kept one in the folds of her dress, interpreting the direction the hare ran on release as an augury for her success against the Romans in the 61 CE uprising.[2]

The hare slowly adjusted to our presence and raised its ears cautiously before bolting off to join its companion. Our serendipitous encounter had lasted no more than a minute in total but the incident had somehow put a temporal brake on the day. As the hare moved off, time – at least, the quotidian progression of past, present and future that embraces cause and effect – was finally unfrozen.

We continued our walk along a broad swathe of firm shingle and sea thrift that Hanne called the Fairway, which lead us to the edge of a tidal creek close to Halfway House. The highly prized real estate of Blakeney was visible across the channel that separated us from the 'mainland', as was the village's St Nicholas' church high above the houses and to the west of this, the iconic, much-photographed windmill at Cley. In the network of creeks and mudflats that fringed the channel, redshanks were alternating between stabbing for invertebrates and flying short distances, calling plaintively as they went. At the muddy creek margins creeks, a miniature forest of marsh samphire was starting to emerge, although it was still too early to pick.

Returning to the car park, we walked along the sloping beach next to the outgoing tide. Beyond the silhouetted fishing boats ahead, the cliffs of West Runton were visible in the distance. Some of the pebbles beneath our feet would probably have originated from there – flint nodules fallen from the eroding cliffs that had been broken and polished in the churning tide before being dragged here by longshore drift. In the dip of the beach, out of sight of the land, all we could hear was the wash of waves on pebbles and the muffled clatter of an unseen skylark. This was the enduring music of this place: the same sound that fell upon the ears of Victorian lobstermen and medieval sailors. The sound heard by Palaeolithic hunters and, possibly, long before this, by bands of *Homo antecessor* foraging the shoreline. A sound unchanged, with the exception of the intermittent hum of traffic that carried on the wind from the coast road, a recent addition to the chorus. The music of pebbles on the seashore, an elemental noise: flint music.

*

The longest spit in England lies sixty miles south of Blakeney Point in neighbouring Suffolk. This one, Orford Ness, has the all the characteristics of an island and the sense of isolation that goes with that. Orford Ness is not an island, however, but a ten-mile stretch of shingle that begins close to the coastal town of Aldeburgh, a genteel salt-washed resort best known for its annual music festival and Benjamin Britten operatic connections.

A spit that resembles an island yet is called a 'ness' – an Anglo-Saxon word for 'nose' that describes a headland or promontory – Orford Ness is a luminous landscape of shingle, birds and secrecy. A former top secret weapons testing site, now under the auspices of the National Trust, it continues to exude an air of secrecy sufficient to make even the modern-day visitor feel as if that they are standing on forbidden territory. This was once the case. Between 1913 and the mid 1980s, the site was firmly out of bounds to the public, an incontestably no-go zone where all manner of military testing and radar research took place. Its former exclusion from the public gaze is now part of its appeal but, even without this, Orford Ness is a highly evocative sort of place. In recent years, the spit's unique combination of dark history and melancholy landscape has resulted in it becoming a holy ground for a particularly niche variety of art and literature. Drama performances and art projects regularly take place here. Choirs sometimes perform in the remaining military buildings, and numerous films and sound recordings have been made on this island-like territory. All have tried to tap into the Ness's peculiar genius loci.

The German writer and academic W.G. Sebald, in his book *The Rings of Saturn,* describes a 1992 visit to Orford in which, looking across to the pagodas of what he calls 'the island', he likens it to a 'a penal colony in the Far East'.[3] Sebald's fiction, concerned as it is with history, identity and loss of memory, belongs to its own unique genre that might best be termed 'Sebaldian'. Orford Ness, the perfect dark history backdrop for such literal themes to coalesce, might be thought of as the ultimate Sebaldian landscape. And if Orford Ness describes no more than a geographical location, then 'orfordness' might be a suitable coinage for the state of mind engendered by it. Even so, in daylight hours at least, it is usually far from deserted. Orford Ness is the sort of place where, despite its perceived isolation, you are almost guaranteed to bump into day-tripping flâneurs, psychogeographers and Cold War-fixated doom strollers.

The way to reach Orford Ness is by the National Trust–operated ferry from Orford Quay, a short trip across the yacht-bobbing waters of the River Alde that enhances the island ambience of the place. Once across, there are signs that point out suggested routes around the ness, as well as more sinister ones that warn of unexploded ordnance. Most of the detritus of its secret military past has long been removed, but here and there lying on the shingle amidst clumps of sea kale and teasel are tangles of tortured metal and scraps of wire netting. It is hard to determine what purpose they might once have served. The red-banded lighthouse, once a focal point on the far side of the spit, is now gone. Battered by storms and undercut by the relentless erosion of drift-shifted pebbles, the lighthouse was in imminent danger of collapse and so was proactively pulled down in 2020, its concrete rubble left to mingle with the sea-smoothed pebbles on the beach.

I travelled to Orford Ness with friends on a windy day early in the season, driving to Orford Quay before taking the ferry across the river. Access was more restricted than it had been on my previous visits. Now, not only was the lighthouse gone but the site itself and the nearby shore were out of bounds for reasons of health and safety. The most iconic structures that remained on the Ness were the distant bomb-testing 'pagodas', which could only be visited on an occasional guided tour. Less than the Far Eastern penal colony imagined by Sebald, their distant silhouetted form reminded me more of Neolithic long barrows: passage tombs stripped of their skeletons and earth cover: structures built to stress-test weapons of death that channelled five thousand-year-old chambers for the dead. More prosaically, the pagodas brought to mind concrete bus shelters.

Closer was the dark wooden tower of Black Beacon, a two-storey structure originally built to house an experimental power generator. This was now home to a photographic gallery dedicated to the military history of the ness, while its upper floor hosted a looped recording of poems by the Ukrainian-American poet Ilya Kaminsky:

I come to the shingle spit/to watch/how silence/hammers us, ungrammars us.[4]

The poetry cycle, originally intended to be listened to on headphones whilst traversing the Ness, was commissioned for a 2021 on-site multi-media exhibition entitled *Afterness.* It is undoubtedly an evocative work, fully in tune with the dark atmosphere that pervades the site, yet here was another absence: the poet had never actually managed to visit as the Ness as he was prevented from travelling here because of the ongoing Covid pandemic.[5]

The quiet looped poetry of the upper floor blended seamlessly with the liquid white noise of the sea-washed shingle and the peewit of lapwings rising from the marshes. The quartered windows offered framed views of the corrupted beauty of the spit with its scattered military buildings that stretched away to the south. Scattered here and there, silhouetted like Antony Gormley figures in the harsh coastal light, were our fellow visitors, no more than a dozen all told: small figures in a big landscape, humanity dwarfed by shifting landforms and the waning hubris of grand projects.

Opposite the steps that lead up to the Black Tower's exhibition space was a single-storey building called the Power House. Now it housed as a low-key geology exhibition that gave the lowdown on longshore drift and the shifting of shingle. Its window sills were covered with pebbles and shells along with jetsam that had found its way to the Ness – plastic bottles, aerosol cans, plastic flip-flops. A notice read:

More than 90 per cent of the pebbles on the beach at Orford Ness are made of flint or the chemically similar chert. Most of these pebbles are between 65–144 million years old. Some of them were transported by ancient rivers up to 7 million years ago and deposited to the north of Orford Ness in the sediments of the Westleton Beds at Dunwich. The erosion of the East Anglian coast and longshore drift have brought these pebbles to Orford Ness.

Fragmented though it is, made up of countless millions of small pebbles, it was easy to overlook that here was a flint landscape of great purity – the concrete, brick and iron that give the Ness its post-apocalyptic atmosphere were little more than recently forged intruders.

On the occasions I had been here before, it was the lighthouse that had been the main attraction – a beacon to aim for on the long walk across the Ness. Now that it was gone, its absence seemed somehow to make the place even more desolate than it had previously felt. Across the River Ore on the 'mainland', Orford village itself was experiencing its own absence. The village's most prominent feature, the tall keep of the twelfth-century castle built by Henry II to assert his power in the region, was swathed in scaffolding, lost temporarily from the landscape. But Orford Ness is all about absence of one kind or another. It represents proscribed history; a stage set for Cold War intrigue and covert militarism.

This whole stretch of coastline has its secrets. The Orford Ness shingle is the same sea-shaped flint as that found at the Sizewell power station further up the coast, midway between the resort of Aldeburgh and the lost port of Dunwich. Here, a nuclear alphabet is still being painstakingly written: A for decommissioned but not decontaminated Magnox, B for (radio-) actively working pressurised water reactors, C for the proposed next generation – EPR, French and German design, a rumour of Gulf Arab finance. Sizewell reads like a Dickens ghost story impressed on the landscape (or more appropriately, M. R. James – 'Oh, Whistle, and I'll hum to you, my lad'): Sizewell A – the concrete ghost of nuclear past; Sizewell B – our nuclear present; and Sizewell C – the as yet unconstructed ghost of a nuclear future. Sizewell A is an abstract sculpture of low, grey concrete blocks, a factory without windows. Sizewell B is a massive hangar with a giant white golf ball domed upon its roof. Sizewell C, as it stands, is the chartered future, currently no more than an empty space in the landscape.

Robert Macfarlane has shifted the nuclear spectre from Sizewell to Orford Ness in his book-length prose poem *Ness*. In this, a sort of latter-day apocalyptic version of *Sir Gawain and the Green Knight*, Macfarlane animates the landscape with characters drawn from the wild nature of the place. These 'more than human forms' converge to become Ness, an entity composed of tidal drift, moss and deep time, a being that has flint hag stones for eyes – 'the slow silicate grammar-gather of flecks of entity, creeping together over millennia.'[6]

Nuclear weapon research undoubtedly took place on Orford Ness during the Cold War years. A necessary evil, many would say, as the potential threat of the death of millions was put out of mind in the relentless search for a suitably terrifying deterrent. The chosen strategy was to try to outthink the enemy at their own game. In these gloomy troubled times concerns about nuclear disaster have been superseded by the threat of environmental apocalypse caused by climate change. Conflated, the two make for uncomfortable bedfellows – nuclear-drenched concrete and a rapidly eroding coastline is a discomforting combination.

There was enough time before the last ferry to walk to Laboratory 1 of the AWRE (Atomic Weapons Research Establishment). The gate that led to the pagodas of the AWRE site beyond was locked, and access beyond was strictly out of bounds. The laboratory, now with just the bare bones of an iron girder roof, was one of six buildings where in the 1950s and '60s atomic bombs were subjected to mechanical and vibration testing that mimicked the rigours they might encounter before detonation. Although no nuclear materials were ever believed to have been used here, high explosives were, and the hope was that test accidents could be contained on such an isolated site.[7] Nuclear or not, an ominous atmosphere envelops the very fabric of the buildings. Shingle has been banked up against the outside walls as additional reinforcement, a stone buffer against any experimental mishap, a comingling of the natural and manmade.

The flint pebbles of the spit, having already been shifted along the coast by the power of the sea, have been bulldozed here into a radical reshaping of the landscape. This was the desultory architecture of war. The gated entrance to the laboratory resembled the portal to a baleful underground car park. Inside the building, looking up to the bare brickwork and the skeleton of the roof, the scene was more akin to something out of Andrei Tarkovsky's *Stalker* – a menacing zone of annihilation, a landscape for waking nightmares. A large rectangular test pit of reinforced concrete occupied much of the floor space on the right. Built to replicate the bomb bay of an aircraft, the pit was now brimful with water, opportunist vegetation framing its perimeter as if it were a stretch of neglected urban canal. All that was missing were beer cans and a feral shopping trolley. All this in gentle gentrified Suffolk: not so much 'silly', as the saying goes, but something rather more sinister.

*

South of the Ness in the direction of Felixstowe is the tiny coastal hamlet of Shingle Street. The name is descriptive, yet only partially correct. While the beach is composed entirely of flint shingle, the settlement does not have a single street to its name. Perhaps this irony is a nod to nearby Aldeburgh where half of the old fishing town's medieval streets have been lost to the North Sea, and the once centrally located Tudor Moot Hall now stands, as if nervously, immediately behind the beach. Shifting shingle is the only constant on this ever-eroding stretch of coastline.

Little more than a terrace of former fishermen's cottages, which are nowadays rented out as holiday homes for those with a taste for minimal landscapes, all Shingle Street possesses other than shingle is a telephone box that sees very little use. The hamlet is also the setting for the nation's second most northerly Martello tower, a component of a white elephant coastal defence project that stretches from Suffolk to Sussex and dates from the time of the Napoleonic wars. In *The Rings of Saturn* W.G. Sebald describes Shingle Street accurately if unflatteringly as 'just one wretched row of humble houses and cottages' and a place where had 'never encountered a single human being'.[8]

There is a hidden history at Shingle Street that is just as dark as that of Orford Ness, although mythology is perhaps a more accurate description as the historical truth here is less authenticated than that of the former top secret research site. There are tales of the sea being set alight here during World War II, of an abortive 1940 invasion and of charred bodies of German soldiers being washed up on the beach. There is also a strong suggestion that such stories were propagated as part of a black propaganda campaign developed to encourage US lend-lease cooperation in the war effort.[9] Certainly, there was a lack of reliable witnesses – the entire village population was evacuated in May 1940, so there were no observers to confirm or deny such reports.

Mythology tends to adhere to certain places, and Shingle Street, with its isolation, its bleakness, its groaning shingle that sounds, as W.G. Sebald describes, 'as if the pebbles being dragged back by the waves were talking to each other', maintains an eerie and foreboding atmosphere regardless of whatever the historical truth might be.

But all is not dark or drear. Shingle Street can also be a place of healing. A closer inspection of the shingle on the beach reveals a line of white shells that stretches from the shoreline to the coastguard cottage terrace. The shell line originated in 2015 when two sixty-four-year-old friends Els and Lida sat and chatted on the beach after a long walk together. The women had grown up together in the Netherlands and both were undergoing treatment for breast cancer. As they talked, they arranged a circle of shells around a plant, and when this was complete they began making a line of shells. Over subsequent months of treatment they returned to Shingle Street to continue the work as they met and talked, extending the line of shells sinuously landwards. As the shell line grew in length, both women's cancers went into remission and the completed line, by now 300 metres long and composed of around 20,000 chalk-white whelk shells, serves as a testament to their friendship and shared recovery. The women still return every six months or so to repair any breaks. Although by nature frail and transitory, the line remains clear and distinct. A trace in the landscape like a surgical scar, it exists as a symbol of hope and healing.[9]

*

On a different coastline is another ness that shares some common ground with the spits and shingle beaches of East Anglia. True to its name, Dungeness on the southwest Kent coast, a headland jutting out into the English Channel a little way east of the Sussex border, does resemble the shape of a nose. As one of the largest expanses of shingle in Europe it is a place prized for both its geomorphology and wildlife. Like Suffolk's Orford Ness – until recently, that is – there is a lighthouse. In fact there are two, an old and a new. There is also a nuclear power station. As in Suffolk, there is an abundance of flint in the form of shingle. Dungeness is unquestionably another place where an overwhelming sense of 'orfordness' can be deeply felt.

The modern mythology of Dungeness precedes it. Much of the location's strange allure is connected with the filmmaker Derek Jarman (1942–1994), who lived here in a fisherman's cottage in the 1990s. Jarman's tenure at this place is immortalised his former home on the shingle, Prospect Cottage, a black weatherboard dwelling where he spent his last years fashioning a garden out of salt-bitten shingle and writing his classic *Modern Nature* as he grew increasingly frail as a result of contracting AIDS. Prospect Cottage sits within sight of Dungeness Power Station, which was operational until 2021, a backdrop of twentieth-century nuclear menace that is still firmly in place in the twenty-first.

Jackie and I arrived on the Kent coast on a searing July day that forecasters were predicting to be the hottest on record for this location. The following day was hotter yet and produced the highest temperature ever recorded in the UK. With a modest degree of coastal cooling the air temperature at nearby Lydd-on-Sea, where we stayed overnight, made it to the 37°C mark, less than the record-breaking high. Without doubt, the shimmering heat helped contribute to the post-apocalyptic atmosphere already associated with this otherworldly stretch of coastline.

The beach, immediately across the road from the B&B, seemed the obvious place to seek refuge. The beach at Lydd is composed entirely of pebbles, almost all of which are flint, mostly black although some are a warm shade of amber. I picked up a piece of black flint that, having virtually no cortex, was almost too hot to hold such was its ability to absorb and retain the sun's energy, which on this hottest of days seemed inexhaustible. The sea itself was as warm and brown as gravy, heated by the incoming tide flowing over sun-baked pebbles. I took a swim, glad of the water's relative coolness, and gave barely a thought to the nuclear power station that was undergoing decommissioning a mile or so along the shoreline (why worry anyway? – I had heard that some Suffolk locals deliberately chose to swim on the beach below Sizewell B because the water was always invariably a little warmer there). Across the bay, Folkestone was clearly visible as a parade of white low-rise buildings. When the haze cleared a little, I could also make out Boulogne gleaming across the Channel. Observing the two ports, without spectacles, from my low point in the water, they were little more than a pale smudge on the horizon, indistinguishable in terms of either geology or human endeavour. All nationality voided, they were as stateless as those ill-fated individuals who attempted to cross these busy international waters on unseaworthy contraptions of rubber and air.

We drove to Dungeness the following morning, parking on the road close to Prospect Cottage. The house, black-painted wood with bright yellow window frames, looked to be well cared for and in excellent condition. A note pinned to the back door explained that it had been bought by a consortium and was currently undergoing repair before opening to the public. In truth, it was the garden I was more interested in. An exercise in making the most of limitations, a metaphor for Jarman's last years, it was a temple to flint pebbles and the salt-tolerant flora that would grow in their presence – sea kale, yellow-horned poppy, red valerian, fennel and clumps of tough spiky grasses. *Modern Nature* speaks lovingly of Jarman's garden work, detailing the triumphs of the challenge he had set himself – how the dog rose has 'put out strong new shoots' and the way the purple sage 'doubles in size every few weeks' – and the setbacks, including his 'sadly scorched' roses. The nearby power station is a mostly an unspoken presence, only occasionally warranting a mention as, for example, it loomed through the sea mist.

The garden may have been more elaborate when Jarman was in residence but even so its spirit lives on triumphantly. Unlike most gardens, the Prospect Cottage garden seemed more an extension of the landscape rather than any sort of imposition on it; a case of nature being gently rearranged, its elements identified, gathered together and sensitively curated. Beach debris, like rusted metal, lumps of sea-bleached wood, bottles and corks, had been utilised to provide makeshift statuary. Poking out of the shingle, circles of larger flint pebbles were arranged like small henges. It reminded me a little of some Japanese Zen gardens I had seen, although here at Dungeness the simplicity and minimalism was dictated more by environment than philosophical outlook.

We drove on a little further past the red and white banded new lighthouse to a pub close to the old lighthouse and the power station – the Britannia Inn ('Fish & Chips, Pizza') – that had trestle tables lined up outside that afforded unbroken views to the concrete edifices of Dungeness B. It was hard to imagine something that could be simultaneously both so English and so weirdly incongruous. A few old railway carriages stood alongside the roadside. Some claimed to be artist studios although each was closed, uninterested in touting for business in the heat. Signs in the shingle warned of the necessity of a licence for filming and photography in specific areas. The location has become commoditised; the view, privatised. Like Orford Ness to some extent, the weirdness of Dungeness has become a brand with associated commercial interests. Loyalty cards, whatever their psychogeographical affiliation, were invalid here.

A boardwalk led across the shingle to a bench. Coming back in our direction towards the road were two policeman carrying binoculars. We had already noticed a large police presence in the area – patrol cars, transit vans with anti-riot shields poised above their windscreens. At first, naively, I had thought it was simply for security, keeping an eye on the power station, an obvious if not particularly vulnerable target for would-be terrorists. Then it dawned that they were here to watch the sea for migrant boats. France was close enough for the port of Boulogne to be visible; the English Channel was about as calm as it gets. It was high season for small boat crossings.

Some of the houses that lined Marine Drive in Lydd had first floor balconies that looked out to sea. A few had flagpoles with flapping St George or Union flags. Here at England's south-eastern edge, the Continental 'other' in plain view, expressions of nationalism appeared to be full-throated. It was the sort of place a right wing nationalist politician might choose for a photo shoot, exasperation written on his face as he pointed across the Channel to some imagined foe. I wondered what sort of reception any migrant who successfully beached here would receive from those who had watched their approach through the telescopes on their verandas. Somehow I doubted that many would have the kettle on, tea and biscuits ready in welcome.

Dungeness's watchfulness is nothing new. It is a historical fact, dictated by location and landscape, that Dungeness has long been keeping its ears and eyes open to intruders from across the sea. In the heat of the first afternoon I had taken a walk to see the Denge sound mirrors located on an island in a recently created RSPB reserve. Now designated Scheduled Ancient Monuments, the sound mirrors, constructed of concrete in three radically different designs, had been built between 1928–1935 and were an intriguing precursor to the invention of radar just before World War II. The idea was that they would detect the sound emanated from low-flying enemy aircraft coming across the channel – an early warning system of 'Listening Ears' as they became known. Strange objects to find in any landscape, let alone here among the shingle and marshes of the West Kent coast, two are shaped like Jodrell Bank-style parabolic reflectors, while another much larger one resembles a curved dam wall.

The sound mirrors are just another facet of the strangeness of this stretch of coastline. If liminal places like this were ever to participate in a twinning scheme, then Orford Ness, with its secret bomb-testing facilities and 'edge of the world' atmosphere, would be an obvious contender. Places shaped from flint, sea and salt air, both Dungeness and Orford Ness watch and listen as the pebbles grind and roll on their beaches, ever shifting in Heraclitean flux. *Panta rhei*: ever moving; never the same, always the same.

*

The following day, as we drove west into East Sussex, the heat finally defeated us. The plan had been to visit some of the Romney Marsh churches. We managed a couple, and succeeded in finding Jarman's plain black gravestone in the churchyard of Old Romney Church, but the stifling heat put paid to more extensive sightseeing so we returned to the coast. At Rye Harbour we came across a group of grey seals slumped on the mud bank of the river estuary. The animals looked disconsolate, confused as to what to do in the unnaturally hot weather. Even the claggy grey mud they languished in seemed unable to offer any respite from the heat.

Rye Harbour was where the blue flint used for high quality pottery had once been gathered to be transported by boat and train to the workshops of The Potteries. On arrival in Stoke-on-Trent the flint was 'calcined' (combined with coal and heated), before being ground into a powder that was employed to improve the strength and quality of finish of pottery. The addition of a good proportion of powdered flint proved helpful in preventing both the crazing of glaze and the warping of tiles, although an unfortunate consequence was that the dry grinding that took place caused silicosis. In response to this in 1726, Thomas Benson patented a method of grinding the calcined flint underwater.[10]

Rye flints in the form of small blue pebbles – 'blue boulders' – were collected from the foreshore by local 'boulderers', who piled them up in heaps for loading onto boats that would be floated on the tide into Rye Harbour. Here they would be unloaded into waiting railway trucks destined for The Potteries. Some flint was also transported by sea to the ports of Runcorn and Selby, from where it would travel the final leg by rail. The export of Rye blue boulders continued until the 1950s when the industry went into terminal decline for various economic reasons. In 1960 the railhead at the harbour was finally closed.[11]

It was supposedly possible at low tide to see small piles of blue flints on the shore that had been collected but never moved from the harbour. But it was several hours from low-tide and much too hot to linger, so a cursory exploration from the car park was all we could manage before succumbing to ice creams from a mini market. We returned to the car to drive west through the sun-bitten countryside of the East Sussex coast, joining a frantic two-way cavalcade of traffic clasped between chalk downs and a near-continuous urban strip where the delirium provoked by dreams of escape seemed quite tangible. More extraordinary heat, more chalk, more flint.

FIVE

CAUGHT KNAPPING

Happisburgh, Lynford – Norfolk | Islay – Inner Hebrides
Coneybury Hill – Wiltshire

We're just a few flint strikes ahead of everything else, that's it.
Samantha Harvey *Orbital*

One day a man was out walking his dog on a beach at Happisburgh (pronounced 'Hazebruh') on the northeast Norfolk coast. The year was 2000, and the new millennium had begun without any sign of the feared digital 'millennium bug' Armageddon having taken place. Walking close to the crumbling cliffs that distinguish this ever-diminishing coastline, the man, Mike Chambers, noticed something odd jutting out of the boulder clay. Little was visible, but to Chambers' eye it looked like the tip of a Stone Age hand axe. His instinct was correct. Although he did not immediately realise the full importance of his find, his wife convinced him that he should take the axe to a museum for identification. It was well that he did. This chance discovery turned out to be an object that would challenge the archaeological status quo and result in a significant redrawing of the established timeline of human evolution.

Because the hand axe was found in situ, protruding from datable Cromer Forest Bed sediments exposed at low tide, it was possible to date the object with considerable accuracy. Subsequent investigation revealed the hand axe to be at least 850,000 years old, the oldest of its kind ever to be discovered in northwest Europe. Previous to this, the earliest known human occupation of Britain, dated to around half a million years ago, was thought to be that of *Homo heidelbergensis* as evidenced by fossilised bones and flints found at a site at Boxgrove in West Sussex.

The Happisburgh Hand Axe, as it came to be known, is an object of indisputable beauty. Created from shiny black flint, its delicate tear-drop shape could only have been achieved by precise and skilful knapping. It was also clearly a highly practical tool that could have been used for any number of scraping, cutting and butchering tasks. Effectively, it was a Swiss army knife for its time. The axe was most likely the handiwork of *Homo antecessor*: so-called 'Pioneer Man', a distant ancestor of modern man. Given the confidence with which it can be dated, the axe has proved to be a remarkable find – a single artefact that has turned prehistory on its head. Hitherto, it had been thought that Britain was too cold to be inhabited by hominin species at this time. This clearly was not the case.

A more recent, closely linked, find at Happisburgh, discovered while carrying out a geophysical survey in 2013, has been of footprints of an early human foraging party preserved in the sediment. These revelatory footprints, temporarily exposed by beach scouring following a storm, have proved to be the earliest human footprints found outside Africa. Like the hand axe found in the cliff a decade earlier, these also appear to be evidence of *Homo antecessor*.

When the footprints were first impressed in the mud, the environment would have been significantly different from that which exists today. At that time Britain was still firmly attached to the European mainland by a wide land bridge, and the site of what is now the coastal village of Happisburgh was located inland on the estuary of the ancestral River Thames, which then followed a more northerly route to reach the sea (subsequent glaciations would eventually push the course of the river further south). What was remarkable was that the evidence of a muddy walk that took place more than 800,000 years ago could be revealed in such a serendipitous manner. Against all the odds, the footprints provided an extraordinary snapshot of an event that took place in deep time. The footprints have been interpreted as belonging to an extended family group – adults, children and perhaps even grandparents. The walk took place in a climate that might resemble that of southern Norway today. Pine forest cloaked much of the land at that time, and elk, beavers, horses and hyenas would have roamed freely. As did mammoths – this wandering band of early humans would have been near contemporaries of the West Runton Elephant, the large steppe mammoth whose fossilised bones were discovered further along the coast close to Cromer.

Before achieving fame for its time-bending prehistoric discoveries, Happisburgh has long been known for its savage coastal erosion. Local television news often reported on the village's treacherous erosion as field reporters gave harrowing accounts of houses that had ended up perched precariously on the edge of cliffs. Despite various preventative measures being put in place, the North Sea was relentless, continually chipping away at the land. The village caravan site was frequently forced to revise its boundaries, and the cliff edge was approaching ever closer to the lighthouse, a familiar red and white banded icon. It was only a matter of time before this too would be overcome by the waves and lost to the sea.

*

It was probably Happisburgh's prehistory rather than its reputation for erosion that made it the location of choice for the flint-knapping course I attended on a hot day in the summer of 2021, the second year of the Covid pandemic. The course, entitled *Deep History Detectives*, was run by Dr James Dilley, an expert in prehistoric tool production.

We sat outside under an awning, the gentlest of breezes blowing through to cool us, the neighbouring meadow humming with a high summer soundtrack of bees methodically working its flowers. The others on the course – all men – were around the same age as me and with similar proclivities and interests. One confessed that it was the first time he had really been out since the outbreak of the pandemic the previous spring. After obligatory health and safety instructions we were given a demonstration on how to knap a piece of flint into something that vaguely resembled a Palaeolithic hand axe. The process seemed counterintuitive in some ways – the need to strike the flint in a certain place in order to make it sheer and flake elsewhere on the core. James encouraged us to try and get into the mindset of the axe maker. 'You have to override modern aesthetic instincts and try and engage the Palaeolithic part of the brain… it's something we all have.' The emphasis was on producing a workable tool rather than something that was merely aesthetically pleasing. The exercise, beyond its practical aspects, was an attempt to make a connection with the deep past and the fundaments of what it meant to be human. For me at least, it was hard to see such things beyond their abstract beauty, as there was little if anything to draw upon to in respect of the human qualities of the individuals who had created such objects many millennia ago.

As we diligently knapped away at our cores, willing them to transform miraculously into axes, a midden of flint flakes built up on the plastic sheet at our feet. Waste products like these, rather than the tools themselves, were the most commonly found prehistoric objects. It took a trained eye to identify them as the residue of human handiwork and separate them from an untold number of flint shards that were simply the result of ploughing or frost action.

I was quietly pleased with the axe I eventually managed to make, although I knew that despite a semblance of form that was vaguely Palaeolithic it would be of little practical use in butchering an animal, or even cutting meat. James was encouraging, nevertheless, reminding us that he had been making such tools since childhood and that the skill was not something that could be picked up without extensive practice. In truth, my own modest effort resembled something made by a child loyally copying their father: an object made with enthusiasm rather than experience.

Our instructor's skill was plain to see. He handled and turned the flint with consummate ease as the resulting tool took shape. What was most intriguing was when he came to demonstrate the Levallois method that was used to produce the super-sharp tool favoured by Neanderthals in Europe between 300,000 and 50,000 years ago. This required a more sophisticated technique that involved the creation of a striking platform at the base of the core once the stone had been trimmed all around its edges. A carefully weighted blow to the striking platform was then applied to create a fracture that produced a blade-like shard. This, James admitted, was a hit and miss procedure that was not always successful. Certainly, it went well beyond our own limited abilities. But it worked in this case, and after carefully preparing the striking platform, a single deft strike resulted in a perfect Levalloisian-type flake fracturing cleanly away from the core. James held the flake up to the light to show the translucence of its edge. 'That edge is razor sharp. Actually, it is sharper, maybe a just a few molecules thick. You could slice through anything with that.'

The conversation turned to the probability of finding prehistoric artefacts. Flint tools from the Palaeolithic period – or at least, the microliths that were knapped in quantity from a single flint core – were relatively common we learned. Almost as if to prove the point, James's partner had discovered a hand axe on Happisburgh beach that same morning. In areas like Norfolk, where workable flint was plentiful, such items would have been considered as disposable. A single human might produce hundreds if not thousands of worked flints in a lifetime. In areas where the rock was abundant, a flint tool was not worth carrying far when another could be easily and quickly made in situ when it was required. If anything, items like the red deer antlers that were used as a soft hammer for flint knapping would have been considered more valuable and harder to replace. Despite there being only a small global population in the Palaeolithic, if hand axes were made, say, once a week, and then thrown away when they we past their best, it still amounts to a huge number of discarded tools accumulating over a period of half a million years or more.

We were joined by a celebrity guest for the afternoon session: none less than Mike Chambers, the man who twenty years earlier had discovered the Happisburgh hand axe while walking his dog. 'I just saw the point of it sticking out of the cliff and I decided to investigate further,' he explained modestly. The session, which was held inside the village hall and run by Dr Claire Harris, a Palaeolithic archaeologist at London's Queen Mary University and a co-investigator on the Deep History Detectives Project, was concerned with the identification of flint tools and flakes by looking at clues as to whether they were of manmade origin or not. Bags of flakes from the morning session had been brought along for the purpose. The main thing to look out for was evidence of striking platforms and the cone-like fractures known as 'bulbs of percussion', both of which were reliable indicators of human input. Frustratingly, this was not as straightforward as I might have hoped and I struggled to even recognise flakes that I had probably created myself in the earlier knapping session.

In the lunch break between the sessions I took the opportunity to take a look at the display of James's own hand-worked tools he had brought along. He has his own business and an online shop where he sells his work to private collectors and for museum displays and TV props.[1] There was a fine array of hand axes of different shapes and sizes on show, as well as a finely knapped axe made from glass to demonstrate how that manmade variant of quartz possessed physical properties that were very similar to flint. The replicas were all fine examples of the knapper's craft, but there was no attempt at deception – the objects on show were honest reproduction pieces made in the modern day using the technology of the distant past. They served as a tribute to the prehistoric skill set and made no false claim to antiquity.

The replica business has not always been so honest. In the early Victorian era, when it became accepted that artefacts like flint axes were of human origin rather than 'elf-shot' or 'thunderstones' from the heavens, the demand for such objects became considerable. One man above all stepped in to satisfy this demand, offering convincing counterfeits that, although clearly the work of a skilled practitioner, were of no greater antiquity than the cheddar maturing in the pantry. Flint Jack, as he was popularly known, went under a variety of guises – Fossil Willy, Bag o' Bones and Cockney Bill among others – but his real name was Edward Simpson.

Born in 1815 in North Yorkshire, Simpson started out on the straight and narrow archaeological path in the region of Whitby on the Yorkshire coast before succumbing in 1843 to the realisation that it was easier to manufacture flint arrowheads and axes himself than to spend long hours seeking them out in the field. As Flint Jack, he may have been dishonest, criminal even, but he was undoubtedly skilled in his craft, capable of producing fifty arrowheads in a day.[2] Throughout a productive career Flint Jack manufactured a great variety of flint artefacts – arrowheads, axes, spear tips, scrapers – that found their way into museum collections. Many of these remained completely unchallenged in terms of their authenticity for years. It was Flint Jack's increasing notoriety for all manner of archaeological forgery coupled with an intemperate fondness for alcohol that finally saw his inevitable downfall and a spell in prison. Even so, his skill with stone, unlike his honesty, was unimpeachable and it is almost certain that even today Flint Jack's faux handiwork still adorns a dusty cabinet in a regional museum or two.

*

Working with flint to produce a sharp-edged implement is as old as humanity itself; older, in fact, if we consider that we only became fully human when modern man, *Homo sapiens*, emerged out of Africa around 200,000 years ago. Tool-making is one of those things that make us human, as is the ability to control fire. What is perhaps most fascinating about this most ancient of technologies is the way that tool design evolved, slowly, through the Palaeolithic period.

Stone tools are the deep past's calling cards: lithic reminders of earlier human presence. In East Africa where it all began, the simple Oldowan tools found at the sites of early hominins are the oldest to be found anywhere in the world. These earliest stone tools, which take their name from Olduvai Gorge in Tanzania, were mostly simple tools made from river pebbles that had been struck by a hammerstone to create an edge and sharp tip. Examples of these have been found in Africa that date from up to 2.6 million years ago, the handiwork of a distant ancestor, either *Australopithecus garhi* or *Homo habilis.* This technology was eventually taken out of Africa by a later hominin, *Homo erectus*, as far east as China and Java. The more refined Acheulean method that eventually took over from this stone tool industry emerged around 1.7 million years ago. Two sites associated with *Homo erectus* in the Afar region of Ethiopia that date from up to 1.5 million years ago have revealed the presence of both Oldowan and Acheulean tools, suggesting a gradual replacement of one technology for another.

The hand axe discovered at Happisburgh was a representative of the so-called Mode II or Acheulean type, a pear-shaped form that takes its name from the site of Saint-Acheul in France and is associated with *Homo erectus* and later *Homo heidelbergensis.* Such tools were widely produced during the Lower and early Middle Palaeolithic across much of Africa, as well as West, South and East Asia and Europe. The classic shape, which might be said to belong to the 'if it ain't broke, don't fix it' school of design, persisted until around 160,000 years ago, by which time Neanderthals had made the transition to what archaeologists refer to as Mousterian technology. By this time, the Levallois technique, which I had seen demonstrated on my Happisburgh knapping course, had already been in full swing for 100,000 years or so. Once again, this takes its name from a French location: the Levallois-Perret suburb of Paris, where the first flints of its kind were discovered in the nineteenth century. The Levallois technique provided greater control over the flakes produced, which could be used as scrapers, knives or even projectile points. Superior to the simpler Acheulean method, it raised the level of lithic technology to what scientists called Mode III or Mousterian Industry, which takes its name from another French site, Le Moustier.[3]

As Palaeolithic archaeologist Dr Rebecca Wragg Sykes observes in *Kindred*, her tour de force on Neanderthal life, stone tools connected every aspect of the prehistoric world. More broadly, they could be said to be the fundamental units by which everyday prehistoric life can be interpreted.[4] Neanderthals have, until recently at least, had something of a bad press. The popular image of a stooping, thick-skulled ape-person is closely allied to the widespread use of the term 'Neanderthal' as a pejorative word to describe violent behaviour and knuckle-dragging stupidity. Although it is hard to be certain, given the evidence at hand such a characterisation seems both unfair and inaccurate.

Certainly, Neanderthals were physically different: they were heavier and stronger than the *Homo sapiens* they coexisted and interbred with. The scope of their mental faculty remains uncertain, as does the reason for their sudden decline and subsequent extinction, although factors such as competitive replacement, climate change or simply being bred into extinction and assimilated into the modern human genome might all be significant. The fact of it is that, apart from sub-Saharan Africa where the proportion is minute, each of us still carries between 1–4% Neanderthal genes within us. Neanderthals are, quite literally, part of our DNA.

Neanderthals, who share the same *Homo heidelbergensis* ancestry as modern humans, also probably interbred with the mysterious Denisovans, another extinct hominin species of which very little paleontological evidence has yet been found other than a fingertip and three teeth in a cave in south-central Siberia, and more recently a partial fossil jaw from Tibet.[5] Modern humans also appear to have bred with Denisovans and Denisovan DNA is present in Aboriginal Australian, Polynesian and Fijian populations.

The writer William Golding no doubt had Neanderthals in mind when he wrote his novel *The Inheritors.* The primitive Stone Age folk – 'the people' – at the heart of his book are portrayed as not-quite-human, almost ape-like scavengers, Strong in limb but weak in thinking, they have a frailty that sees them inevitably decline in the face of the culturally and mentally superior modern humans (*homo sapiens*) – 'the new people' – that they eventually encounter. Golding can be forgiven for any palaeoarchaeological inaccuracies in the book, as far less was known about Neanderthal material culture in the 1950s when he wrote it.[6] Whatever their shortcomings, Golding is at pains to depict the Neanderthals as compassionate and gentle in demeanour, and perhaps the ultimate message that comes across is that material progress comes with a price – the superior 'new people', with their more advanced language, powers of reasoning and technology appear violent, warlike and ruthless by comparison with the kin of Lok, the trusting but dim-witted central character.

Neanderthal technology was considerably more advanced than their fictional portrayal might suggest. As well as sophisticated stone tools, they were able to build hearths, make adhesive from birch bark, cook and store food and craft simple clothing. They were undoubtedly also able to control fire and evidence points to them having the means of creating it by the sparking of flint. The range of stone artefacts collected from Neanderthal sites is varied – in addition to Levallois flakes, many types of blades, bi-faces and bladelets are composite tools like axes that combine stone and wood. As Rebecca Wragg Sykes asserts: 'Stone tools were the atoms of Neanderthal life.'[7] (By stone, she means flint.)

Neanderthals, who walked the Earth as recently as 40,000 years ago, ranged from Central Asia to as far west as Spain and southern Britain. One Neanderthal site, dated to the Middle Pleistocene of around 60,000 years ago, has been found at Lynford Water, a former quarry, in the Norfolk Brecks. There is clear continuity of use of the flint-bearing landscape here, as the site lies close to Palaeolithic tool sites at Weeting in the Little Ouse Valley as well as later Neolithic flint mines at Grime's Graves. It is also close to the far more recent flint mines at Lingheath and Session Heath, which provided the raw material for the once thriving flint-knapping industry at nearby Brandon just over the Suffolk border.[8]

The site lies just south of the River Wissey and the village of Ickburgh, close to the Danger Area marked on OS maps, a large tract of heath land set aside for military training purposes. In 2002, a large number of *bout coupé* hand axes typical of Neanderthal Mousterian Industry were found in organic mud deposits here alongside the skeletal remains of up to eleven woolly mammoths. The butchery site appears to be around 60,000 years old, dating from the Devensian cold period, a period cooler than today but sufficiently warm for the more cold-adapted Neanderthals to tolerate. There is no definite proof that the animals were hunted and killed by the Neanderthals, so it may have been simply that the carcasses were scavenged and their limbs removed to eat away from where they fell. It seems probable that Neanderthals would have had a campsite nearby, although nothing has yet been found.

Less than a kilometre from Lynford Water is Lynford Hall, which has an arboretum much favoured by birders on the look-out for scarce woodland species like hawfinches. Tucked away in the woods south of here is the Victorian Catholic church of our Lady of Consolation and St Stephen. Once a private chapel for Mrs Yolande Lyne Stephens of Lynford Hall, it is a curious structure with a small round tower capped with a dunce's cap that brings to mind a minaret. The trees that surround the church are a later addition, planted by an Anglican who thought the chapel an eyesore. Naturally enough, the church is built predominantly of knapped black flint. To build with anything else in such a flint-rich region would appear quite unseemly.[9]

*

What is now generally regarded as the Mesolithic began in the post-glacial period that followed the Last Glacial Maximum, which ended in Europe around 11,500 years ago at the beginning of the current geological epoch, the Holocene. Many believe that this has since been succeeded by a new, human activity-influenced epoch – the Anthropocene. After millennia of being covered in ice and too cold for habitation the British Isles started to become habitable once more. It was a time of wandering hunter-gatherers, who made the most of the rich food supply available to them in the newly thawed landscape. Britain was still attached to Europe at this time and small bands would wander across the land bridge of Doggerland to make the most of the rich hunting this virgin territory provided. Around 8,000 years ago, sea levels rose sufficiently for Britain to become an island once more. Those already settled on these shores would continue their hunter-gathering lifestyle in isolation until the arrival of Neolithic farmers around 3400 BCE.

The Mesolithic toolkit mirrors the lifestyle these people followed and consists largely of microliths, the small knapped flint blades used to create composite devices like arrows, spears and adzes. Microliths were produced from bladelets, knapped in number from a common core, which meant that a single piece of flint was capable of producing an array of useful tools. Good quality flint was preferred but where there was not a ready supply an inferior source would be used instead. The process was simple. First the cortex was removed as completely as possible before a striking platform was created on one side, then bladelets were removed from the core by means of a soft hammer or punch. It is this lack of cortex and use of a soft hammer that makes Mesolithic flintwork so distinctive, producing microliths with distinctive abrasion marks and bulbs of percussion that result from the preparation of a platform prior to their removal from the core.

In southern England most Mesolithic sites tend to be found along river valleys, and many reveal the widespread use of flint microliths for use as tools and hunting weaponry. As they were semi-nomadic hunter-gatherers frequently on the move, the sites were places where hunter-gathering bands would have returned to periodically to butcher animals or make stone tools. Often they are associated with middens of animal bones and mollusc shells, along with the remains of foraged foods like hazelnuts that appear to have been an important element of the Mesolithic diet.

In Norfolk, Mesolithic worked flints have been found at Hockham Mere, one of several small lakes in the Brecks that were formed around 10,000 years ago following a collapse of underlying chalk bedrock. The mere has since dried up to leave swampy area called Cranberry Rough, but when it was first formed in the last post-glacial period its hunting and fishing potential would have been readily exploited by Mesolithic hunters, as was the plentiful supply of quality flint for tool making that had been similarly exploited by Neanderthals roaming the same territory 50,000 years earlier.

Kelling Heath, close to the North Sea coast, a glacial moraine formed by the retreating glaciers of the last ice age, is another Norfolk location where Mesolithic flints have been found scattered over an extensive area. These days the heath is the site of large holiday camp and crossed by the heritage railway line that connects the towns of Sheringham and Holt. Its light sandy soil and distinctive fauna of birch and bracken also makes it an excellent place for adders and for elusive nightjars that whir through the air at dusk on midsummer evenings. Ignoring the trees that grow there now, it is easy to see why such a site would have held appeal for hunter-gatherers. The heath's relatively high vantage point would have provided an excellent view of animal movement over the flat plains of Doggerland that in the Mesolithic period lay beyond what is now the North Norfolk shore. Sometimes it is useful not only to try to see through prehistoric eyes but also to actually summon to mind the landscape they lived in; to re-imagine a time when Britain was still attached to Europe, when familiar rivers followed different courses and, rather than a cold dark sea filled with the slow-turning blades of off-shore wind farms, a vast flat plain stretched north and east as far as the eye could see.

*

Scotland, then as now, was different. Other than as beach pebbles along a small part of the coast, surface flint is a scarce commodity. Almost all of Scotland's Cretaceous deposits were eroded during the Tertiary and Quaternary periods. Consequently, where flint does exist it is usually buried under strata of different rock.

One of the few places in Scotland where flint is reasonably accessible is at the Den of Bodham in Aberdeenshire, close to Peterhead, where there is evidence of it having once been mined quite extensively in the Neolithic period. The Den of Bodham flint lies within a chalk deposit on a ridge of high ground, and a total of over 450 shallow circular pits have been found scattered over an area of eight hectares.[10] The flint here is mostly grey-coloured and in the form of pebbles buried beneath glacial till and gravel. Although there are extensive scatters of flint flakes, all the available archaeological evidence points towards no finished implements being completed on site and so the final knapping process most likely took place elsewhere.

While Scotland has a wealth of Neolithic monuments, there is little or no evidence of earlier Lower or Middle Palaeolithic settlement. There is, however, plenty that points to widespread Mesolithic occupation. Until fairly recently, the oldest evidence of human occupation in Scotland came from the east coast of the Isle of Islay, where numerous microliths were discovered at Rubha Port an t-Seilich near Port Askaig on the northeast coast when pigs were brought in by local farmers to help clear bracken.[11] Digging down into the soil, the animals inadvertently unearthed a layer of stone tools. An archaeological team from the University of Reading led by Professor Steven Mithen, a regular visitor to the Inner Hebrides, was brought in to investigate. The flints turned out to be mostly Mesolithic flakes left over from the production of tools like knives, projectile points and scrapers.[12]

Islay was unlikely to have been a permanent dwelling place, but rather a summer camp for hunting and foraging. It was also a reliable source of secondary flint in the form of beach pebbles. The Mesolithic discoveries on Islay point to an occupation of the island anywhere between 10,650 and 6,000 years ago, the period between the end of the last ice age and the first arrival of Neolithic farmers. Beneath the debris, at the bottom of a test pit, an arrowhead was also found that had a central projection at its base ('tanged' in archaeological parlance) that would have attached it to a shaft. Such a find suggests an even earlier human presence on the island, one that could be dated to around

12,000 years ago to the time of the final millennium of the last ice age – the period that saw the transition between the Late Upper Palaeolithic and the Mesolithic. Subsequent excavation of a fireplace would reveal more tools of a similar design to the tanged arrowhead, the shape of which suggests that they were made by roaming reindeer hunters of so-called Ahrensburgian culture, who came from the region that is now Scandinavia, northern Germany and the Low Countries. Before the discovery at Rubha Port an t-Seilich, tanged arrow points of similar design had previously been found on the island of Tiree, on Orkney and also on the mainland at Sheildaig, Wester Ross. This, Professor Mithen suggests, points towards Ahrensburgian people, having first crossed northern Doggerland, circling the northern tip of Scotland in skin boats before venturing south along the west coast to eventually reach Islay.

Both flint and quartz were used as raw materials for the tools, and a lack of cortex on the flint found suggests that the initial stages of manufacture had taken place elsewhere. As the only source of flint on Islay are pebbles on some of the island's west coast beaches it is assumed that flint was collected from here and knapped locally before being carried across the island to the camps on the east coast at Rubha Port an t-Seilich. This is corroborated by evidence of initial stages of flint knapping at sites like Coulererach, which lies close to the beaches of northwest Islay such as Kilchiaran, Saliago and Sanaigmore where flint pebbles are plentiful.[13]

The Mesolithic finds on Islay were recently trumped in antiquity by a 2010 find at Howburn Farm near Biggar in South Lanarkshire. Here, a concentration of unusually large flints was discovered in the same field in which Mesolithic and Bronze Age microliths had been found the previous year. These, along with a flint projectile point that had been broken in two parts, were identified as belonging to the Late Upper Palaeolithic period. Such a discovery demonstrated that people were present in

Scotland around 14,000 years ago, considerably earlier than had been previously thought.[14]

While the site at Howburn was given chronological context by the flint tools found scattered there, the stone itself was absent from the locality. There is no flint to be found in situ anywhere in Lanarkshire and the nearest potential source of raw material would have been outcrops that now lie beneath the North Sea on the east coast.

While it is clear that flint was used opportunistically by those who occupied flint-rich locations, the stone was an essential travelling companion for those moving into regions where it was unknown or scarce. While Mesolithic peoples would travel a long way to source flint for their tool making, they would also carry it a similarly long distance to make use of it at their camps. The Howburn folk, who would almost certainly have only used the site as a summer camp, also made tools out of chert, which was reasonably plentiful locally. Even so, they still made good use of the supply of the superior flint that they had brought along with them.

Flint tools dating from the Mesolithic, Neolithic or early Bronze Age have been found elsewhere in western Scotland, particularly on some of the islands of the Inner Hebrides, like Colonsay, Mull and Tiree. Some of the beaches in this region have an abundance of flint pebbles, which would have arrived here carried by ocean currents from either Cretaceous undersea deposits offshore, or from the Antrim coast of Northern Ireland.[15] Geology has no respect for modern national borders.

*

While small groups of Mesolithic hunters were doing their best to keep warm roaming around a recently thawed Britain, far away, in the land that lay beyond the other end of Europe, a revolution was taking place. It was a revolution that changed the way people exploited the land and how they lived together. At the same time as hunter-gatherers were roaming the shores of

Britain, setting up summer camps in hostile yet food-rich territory, their counterparts in what we now call the Near East were tilling soil, planting crops, raising livestock and building the first cities: at Göbekli Tepe in southeast Turkey, Byblos in Lebanon and Tell es-Sultan at Jericho in the West Bank.

The revolution – more a cognitive shift of how best to survive – would soon travel snail-slow westwards like a breeze rippling through a field of corn. Reaching Greece in about 7000 BCE, the so-called Neolithic Expansion moved across Europe at a rate of about a kilometre a year, to finally arrive in northwest Europe and the British Isles around four thousand years later. As well as common characteristics like the domestication of animals and plants, living in communities and the making of pottery, at the heart of this newly adopted mode of living were new ways of working stone. Neolithic: quite literally, a 'new stone' culture. This new way of life required a change of technology: tools that were fit for purpose in this brave new (stone) world.

The transition, revolutionary or not, was gradual. While agriculture and the domestication of livestock grew in importance in the early Neolithic way of life, the old ways of hunting and gathering wild food also continued. Even so, the die was cast and semi-permanent settlement soon became more widespread. The future would see a greater control over nature and people living together in increasingly large groups. Inevitably it would also result in expansion of territory and competition for resources. Sometimes it would lead to conflict.

Mesolithic people had exploited the landscape in a light-touch manner – hunting game, collecting wild plants and roots, making use of naturally occurring materials like flint and wood for tool-making. Neolithic farmers, in contrast, had a more entrenched relationship with the world they lived in. Rather than existing within the confines of its natural bounties, they rearranged the landscape to suit their own requirements, painstakingly clearing trees and lifting stones from the land to create space for crops. Neolithic people would go on to leave their mark in other ways too, heaving unfeasibly large stones considerable distances across country to create markers of territory and centres of ritual. It can seem almost as if the shifting of stones became a compulsion as the social cohesion required by larger groups of people living together become increasingly important. In Mesolithic times, social groupings had rarely been larger than extended families or loosely organised hunting bands. With the arrival of agriculture and animal breeding in the Neolithic, something akin to what we now think of as society was starting to emerge.

The transition to a more farming-based economy was slow nevertheless, and even when the inspiration for it did arrive it took many generations before it became fully established. The shift certainly does not appear to have taken place quickly and, rather than a rapid uptake of new ideas brought by incoming continental farmers, the eventual adoption of settled agriculture was probably more a matter of responding to population increase and the need to better manage available resources.[16]

There is evidence that points to Mesolithic and Neolithic groups co-existing in close proximity for periods of time. A feast that took place at Coneybury Hill in Wiltshire sometime between 3800–3700 BCE appears to have been attended by both Mesolithic hunter-gatherers and newly arrived Neolithic farmers. The hilltop site of the feast was only 1,500 metres from Stonehenge, or rather where Stonehenge would eventually be raised a millennium later. Excavations of a pit at the site have revealed the bones of both wild and domesticated animals – red deer, roe deer and beaver as well as pigs and cattle. It is, of course, impossible to know precisely how the two cultures related to one another but the animal bone evidence seems to suggest reciprocity. It is quite possible that the purpose of the event was to socially lubricate a relationship that might sometimes be fractious. It may even have been the Neolithic equivalent of a 'getting to know you' BBQ, in which potentially antagonistic neighbours wanted to bury the hatchet (or, rather, the flint axe). While disparate in lifestyle and economy, both Neolithic and Mesolithic shared the same rolling Wiltshire landscape and, given both the relatively low population of both cultures at the time, it is unlikely that there was any conflict over land tenure at this point. Notions of land ownership would inevitably come later, as would the concept of private property and the emergence of social hierarchies and authority figures.[17]

The Coneybury Hill feast might even be viewed as some sort of hunter-gatherer last stand: a celebration to mark the end of an era. It certainly coincided with what would be a turning point in social evolution. In the presence of the more sophisticated Neolithic folk, was this an invitation to finally step out of the wildwood, to develop a new sense of propriety towards the landscape that nurtured them? It is easy, although probably naïve, to see this as a leave-taking of an imagined Eden. Even so, there is plenty of evidence to support the view that a hunter-gathering lifestyle was healthier, less stressful and physically easier than a sedentary one of growing grain and herding animals. Hunter-gathers generally lived longer and, not obliged to chew gritty grain for sustenance, they usually had better teeth too. Fertility was another matter: no longer required to constantly move in the search for food, Neolithic women were in a position to have more children, the flip side of which meant a greater demand for resources, for food and land.

While there was undoubtedly a degree of overlap, a distinct dividing line separated the two cultures. This was manifest in a notable difference between the flint technologies they would come to employ. Some of the flint tools of the early Neolithic were not dissimilar to those that had characterised the preceding Mesolithic period – flakes, leaf-shaped arrowheads, various types of scraper and blades were all still produced. What was markedly different was the development during this period of a new type of ground and polished axe.[18] This was as a result of the demands of agriculture that required the widespread clearance of forest. Such a development created a need for an increasing number of axes, and in turn a requirement for a large supply of flint of a suitably high quality. For the first time in human history, it became necessary to mine flint in bulk from underground sources. Along with agriculture, animal husbandry and the first inklings of civilisation, the world's first industry had been born.

SIX

ROCK OF AGES

Grime's Graves – Norfolk | Cissbury Ring – West Sussex
Cuckmere Haven – East Sussex

In the grass; chip of flint, and mite
Of chalk; and the small birds' dung
In splashes of purest white
Edward Thomas *But These Things Also* (1915)

Seen from above, Grime's Graves in the Norfolk Brecks looks like dimpled baize: a clearing of green velvet moonscape surrounded by the dark mass of the densely planted conifers that makes up most of Thetford Forest. The grass-carpeted mounds resemble giant molehills, while the shallow depressions that lie between them give the impression of small lunar craters. These dimples are not the result of meteor strikes but of human activity: the imprint of mine shafts that were active here over a period of several hundred years during the Late Neolithic period.

Although it seems likely that only two or three of the pits were ever operational at any given time, these flint mines were worked more or less continually between 2600–2200 BCE and intermittently during a secondary mining phase in the Bronze Age. Later still, some of the pits were used for inhumation burials in the Iron Age – this is a place where ghosts and flints vie for attention.[1] What remains today is evidence of a mine complex that became redundant as technology advanced, although the demand for quality flint in the Brecks did eventually resurge when it was later required for use as a building material or for strikers for flintlock muskets.

The Grime's Graves mine complex existed at a time of widespread stone shifting: not only the extraction of flint from the ground as seen here, but also the moving and placement of much larger stones for reasons that even now are not fully understood. The operation of these mines in eastern England was contemporaneous with the construction of Stonehenge and the Avebury Ring in what is now the county of Wiltshire in the southwest. The mysterious 40-metre-high mound of Silbury Hill, close to Avebury, was another contemporary project. However they were organised – democratically or otherwise – the grand schemes of the Neolithic never seemed to want for determination of will or lack of ready labour.

The extensive mining site of Grime's Graves covers an area of 7.6 hectares on a low chalk ridge close to the Norfolk–Suffolk border. Altogether, a total of 433 mines and pits have been identified here, twenty-eight of which have been excavated. While almost a millennium later in date than similar flint mine complexes in Sussex, Grime's Graves is easily the largest site of its kind in Britain, although it is smaller than its counterparts in Europe, like Jablines in France and Krzemionki in Poland. It seems likely that when it was a working site the valley here had a stream running through it, a necessary resource for those who camped here to mine the flint. There is no evidence of a permanent settlement, however; Grime's Graves appears always to have been a place of work rather than somewhere to dwell full-time.

The site's existence was down to its abundant supply of high quality flint, a resource that was in great demand in the busy land-clearing days of the Neolithic. Naturally, it was never known as Grime's Graves to those who mined here: any name or identification attributed by the original workers has long been lost to the chalk. The current name arrived much later in Anglo-Saxon times and is a corruption of Grim, an alternative name for the Saxon deity Woden.

A new way of living required a new technology. As before in the hunting-gathering Mesolithic era, flint was central to requirements. The Neolithic revolution that led the way to widespread agriculture required good quality flint in quantity for its wooden-handled axes, the essential tool for clearing woodland. The flint found at the Grime's Graves mines came in two forms: nodules that lay close to the surface and the superior stone that could be found at greater depth in extensive sheets known as tabular flint. Grime's Graves was by no means the first mine of its kind in Britain. The mining that took place here came later than similar mines on the South Downs of West Sussex, where Neolithic people had been excavating flint at locations like Cissbury and Harrow Hill more than a thousand years earlier.

As with the mines in the South Downs, the principal excavation tool of the Grime's Graves miner was the red deer antler pick, a resource that has been found in abundance at excavations. Precisely why the Neolithic people who mined the flint here felt the need to go so far beneath the surface is not entirely clear. While it is true that the best quality material for making axes was the tabular flint that was located at depth, poorer quality surface flint had been used successfully in the Mesolithic period for larger tools like tranchet axes. If it was not purely a matter of quality, then it was probably that the difficulty entailed in reaching the deeper-bedded flint afforded the stone a greater cultural value. There are many indications that the mines were excavated (and closed) in a ritualised way, and that at least some of material extracted was shaped into objects for ceremonial rather than practical use.

The evidence from some of the Grime's Graves shafts suggests a highly ritualised setting in which offerings were made and ceremonies performed. These may have been relatively unproductive shafts that were specifically chosen and set aside for ritual use. A few, like Pit 15, contained curious offerings like carved chalk balls and phalluses, while another, Greenwell's Pit, contained the bones of a carefully buried dog. Most intriguingly, another gallery of Greenwell's Pit contained an arrangement of red deer antlers, a polished greenstone axe and the skull of a wading bird.[2]

The archaeological evidence shows that the majority of the tools made at Grime's Graves were arrowheads, discoidal knives and scrapers. Any axes that originated here appear to have been largely unfinished rough-outs that would be completed elsewhere. Of those axes, often ground and polished like the Cornish axe found in Greenwell's Pit, a number would have been kept solely for ceremonial purposes and not have seen much or any practical use.

Some of the rough-outs would have been abandoned because of an imperfection in the flaking process – a misdirected blow, or the discovery of a flaw in the flint nodule. After making a rough-out, the next stage was to produce a 'preform', in which any remaining cortex was removed, high ridges were reduced in size and the cutting edge was retouched by flake removal to produce a straight and sharp edge.[3] Once the preform had achieved the correct shape further retouching was carried out on the edges to make the axe ready for grinding and polishing. Compared to the knapping preparation, this was a slow, laborious process that involved grinding the shaped flint against an abrasive piece of sandstone known as a *polissoir*. Those axes destined to be used for working wood would then be hafted on to a wooden handle and secured with leather and fibre strapping.

Estimates vary but research suggests that a small axe might require something like five hours work, while a larger one might take up to forty hours to complete.[4]

The obvious question is: why go to all the trouble of grinding and polishing an axe when so much additional labour is required? In practical terms, a polished axe that has a smooth cutting edge is less likely to break from a blow at the wrong angle. There is also the matter of aesthetics: a carefully polished axe is a beautiful thing and an object of prestige. Such a valued item, with its innate investment of time, trouble and human skill, would be considered of higher worth than a more prosaic, run-of-the mill tool. It is this intrinsic value that affords its place as a ritual object. Unsurprisingly, many ground flint axes have been found at locations that were probably considered sacred to Neolithic folk. In their role as votive offerings, many appear to have been intentionally destroyed, often deliberately broken in half. Broken or not, these same ritual objects are often in mint condition, showing no sign of having been used in a practical way before internment. As with any sacrificial practice, the belief appears to be that the greater the value of the gifted object, the greater its efficacy as an offering.

The ritual offerings found in the Grime's Graves pits, as well as in the earlier Sussex mines, suggest that ceremonies associated with mining were commonplace. As with the extra endeavour required for grinding and polishing an axe, the jeopardy of delving deep underground to collect the raw material may have given the flint a greater symbolic value – a precious thing that came from the perilous, secretive world beneath the Earth's surface.

*

I visited Grime's Graves twice within the space of a couple of months. The first occasion was the standard visitor experience – a quick look around the display next to the ticket counter and then, suitably helmeted, a climb down the ladder into the single pit that was open for visitors. The pit was well-lighted and the aluminium ladder that led down to the bottom, reassuringly secure. At the lower level, the narrow side galleries were illuminated by carefully placed electric lights, although solid metal grills prevented further investigation. The electric light was harsh and yellowish and because of this I could not manage to take any decent photographs whichever way I adjusted the settings on my digital camera. I found the overall experience a little disappointing, but there really was not that much to see. Even so, I had hardly expected the full Neolithic experience of going underground with little more than a deer antler and a lighted taper for company.

Walking around the site afterwards, along the warren of paths that connects the mounds and depressions of the pit complex, the most noteworthy things in sight were the sheep – a Shetland variety that seemed unfazed by wandering humans – and the sheer number of rabbits. The Brecks has long been a rabbit-raising region, its infertile sandy soils being poor for agriculture but ideal for rabbits to burrow. The nearby town of Brandon was once notorious for its perennial reek of uncured rabbit pelts. Many of the depressions contained burrows and I checked some of the entrances for worked flints, although it was hard to convince myself that any of the flakes I found were the work of Stone Age tool makers rather than merely fragments broken naturally by frost or plough.

Another notable feature of the site was its near silence, a rare commodity even in rural Norfolk. But we were far from any main road and the only sound, other than the whisper of a breeze through the pines, was the muted clatter of high flying skylarks and a very distant cuckoo.

*

Of the ten known flint mining complexes in Britain, several are found in close proximity in the chalk country of the South Downs in West Sussex. Although smaller in scale, these were all in operation far earlier than Grime's Graves and date back to the early Neolithic period. The mines at Cissbury Ring, close to Worthing, are probably the best known. Cissbury Ring takes its name from the Iron Age hillfort that was constructed here around 250 BCE: a large earthwork of ditch and ramparts that was superimposed on a much earlier flint mining site that was operational between 3900–3660 BCE. What remains today is similar to that which is found at Grime's Graves – mounds and hollows that mark the location of the former pits.

Cissbury Ring was surprisingly quiet when Jackie and I visited on a scorching afternoon in the summer of 2022. No doubt the beach at Worthing was a more appealing option, but the car park at Storrington Rise was almost empty when we arrived; just a few local dog-walkers doing their best to cool down their panting, overheated pets.

We followed the signed footpath uphill through sun-bleached meadows where the thistles were thronged with butterflies like marbled whites and, appropriately, given the location, chalkhill blues. There were few birds to be seen apart from high-flying swifts and wood pigeons that flapped lethargically between the trees, stalling inelegantly in midair as was their habit. It was probably too late in the breeding season to expect to see much but even so, it was depressing to read in a brochure I had picked up at Worthing Museum that wheatears had once been abundant here. Abundant, that is, until Sussex shepherds took up the habit of trapping them to sell for the London restaurant trade. Horrific, but who was really to blame – impoverished shepherds or wealthy metropolitan gastronomes with a taste for small birds? Despite this taint of mean local history, the landscape, all open grassy slopes and scattered trees, looked like perfect wheatear country.

As we climbed the path, the English Channel unfolded beyond the coastal sprawl of Worthing-on-Sea to the south. Wind turbines churned the air in the distant haze. We passed through a thin belt of woodland, and Cissbury Ring came into view: a fence, a ditch, a raised rampart with steps; beyond this, the mound of the hillfort itself. We arrived at the mound's southwest corner where many of the Neolithic pits had been, still plain to see as depressions in the ground overgrown with heather and low shrubs. Fragments of flint and chalk bedrock lay everywhere, scattered on the short-cropped grass and on the path that circled the mound. The flint was pale-coloured, mostly creamy grey. Out of habit, I pocketed a few pieces as we made a circuit of the Ring.

A small group of horses were alternating between grazing the parched grass and seeking refuge beneath one of the few trees available. At intervals along the path were National Trust signs with QR codes that linked to details of the history of Cissbury Ring. Approaching the end of our circuit I was pleased to find a gnarled old bench that bore a plaque with a QR code that linked to a biographical sketch of John Pull, who more than anyone warranted a mention here. While General Pitt-Rivers, founder of the eponymous Oxford museum, had excavated several of the Cissbury pits in the 1860s and '70s, it was local man John Pull who would go on to become the real authority on flint mining both here and across the whole South Downs region.

I had learned a little about John Pull the previous day at Worthing Museum. Pull was a locally born, self-taught archaeologist who did not always received the full credit he deserved. A remarkable man who led an equally remarkable life, his pioneering work in the 1920s on the flint mines located in the South Downs close to his Worthing home would prove to be invaluable for the understanding of mining in the Neolithic period. Unfortunately, because of his lack of formal training, his findings were often disregarded or dismissed outright by the solidly upper class archaeological community who were his contemporaries. Pull's story has parallels to that of Basil Brown, the amateur archaeologist behind the Sutton Hoo ship burial discovery in Suffolk, whose professional ability was similarly looked down on by arrogant Oxbridge academics.

John Pull's life was event-filled from the outset. Aged just seventeen, he enlisted for the 1st Battalion Rifles Brigade in 1916 and was sent to the Western Front. Declining promotion because he refused to order men to their deaths, he was gassed at Mons and later captured and handed over to the care of German nuns at an Ursuline convent – kind women with whom he kept up a lifetime correspondence. When World War I ended Pull returned to his native Sussex to work as a postman. Having a lifelong fascination in both the natural history and archaeology of the Sussex Downs, in 1922 he began excavations at the 6,000 year-old Neolithic flint mines at Blackpatch. He soon found himself marginalised by some members of the archaeological hierarchy, who excluded him from tours and even removed his name from the official 1924 excavation report. Unbowed by such shoddy behaviour, he continued his work at Blackpatch until 1932 and even published a well-regarded book about his findings. In later years he investigated the Neolithic mines at Church Hill and finally, in the 1950s, those at Cissbury Ring. In contrast to the academic and social snobbishness of his detractors, Pull's own excavations were noteworthy for their inclusiveness, welcoming women, working class enthusiasts and ex-servicemen equally.[5] Following a lifetime of dedicated archaeological work in the Sussex Downs, Pull's demise was sudden, violent and ill-fated. He was shot dead by a bank robber in 1960 while working as a security guard at the Durrington branch of Lloyds Bank.

Worthing Museum, which holds his archive, has a permanent display devoted to Pull's work at the Blackpatch site. The same cabinet holds an information board dedicated to the man himself: *John Pull – Worthing's Hero Archaeologist.* Another display, entitled *Legacy*, has a photograph of the immaculately dressed archaeologist holding a deer antler and looking up towards the camera from a shallow pit. The text, quite correctly, describes Pull in the warmest terms as a 'modest and determined man' who is remembered fondly by the children who once worked with him on digs. More revealingly, it pays him the credit of recognising Neolithic miners not as primitive, earth-plundering savages but as folk with 'quite a degree of civilization, including even some knowledge of agriculture'.

*

There is a viewpoint on England's south coast that is familiar to many. Like the Grand Canyon or Niagara Falls it depicts a landscape that is recognisable to even those who have never been there, although in this case they may not be able to put a name to it. Beloved of photographers, it could be said to represent a visual cliché – a readymade postcard or calendar shot. This is the classic view of Cuckmere Haven in East Sussex that has the coastguard cottages at Seaford Head in the foreground and the rolling chalk cliffs known as the Seven Sisters stretching away east in the direction of Eastbourne beyond.

Cuckmere Haven describes the bay where the Cuckmere River flows into the English Channel. The river is a textbook meander, a serpentine thread that, having flowed down from the South Downs, twists its way through the flat coastal floodplain to the sea. There is a classic, although perhaps less well known, image

of this too: a black and white photograph taken by Bill Brandt in the 1930s, a minimalist print reproduced in such high contrast that it appears almost abstract. These days the river's meanders are little more than fluvial ghosts as the main channel of the river has long been canalised to improve water flow.

Jackie and I drove to Cuckmere the day after our visit to Cissbury Ring. The weather had cooled a little and it was a fairly typical July day – warm and slightly hazy with a fecund ripeness in the air. Walking down from the car park of the Golden Galleon pub, the Seven Sisters did not come into view until we almost reached the beach. The shingle beach itself is bisected by the river in its last few metres to the sea. Visitors were milling around on both sides of the channel, but it was too deep and swift-flowing to attempt crossing.

The view east to the towering Seven Sisters was as spectacular as we had anticipated. Pristine white, the cliffs might be made of cake icing, or even perhaps ice itself – a reminder of the great glaciers that once covered the land here. Beyond the undulating sweep of the Seven Sisters (which to our eyes looked more like eight) were the more distant cliffs of Beachy Head. Through binoculars we could make out the flint strata in the chalk. Layered at regular intervals, there might be around seventy altogether between cliff top and beach. Laid down during the Late Cretaceous period, when sea level was around 200 metres higher than it is today, the chalk close to beach level was deposited around 87 million years ago while that close to the cliff top was newer, formed about 84 million years ago. From top to bottom, a sequence that took three or four million years to form, each stratum of flint is a measure of passing years like the tree rings of an ancient redwood. Here was a tangible geological timeline, deep time preserved in chalk and flint.

The beach was composed almost entirely of flint pebbles of every shape and form. We spent a little time looking out for anything unusual, and it did not take long before I found a couple of hag stones – flints that have a naturally formed hole going through them. They are common enough on beaches like this, although not necessarily easy to find. Derek Jarman had collected stones like this from the shingle at Dungeness and threaded them on rope as sculptures for his garden at Prospect Cottage. Considered to bring good fortune, hag stones have long figured strongly in European folklore. One old Norfolk tradition was to hang them up in stables for the magical protection of horses. George Ewart Evans, writing in *The Pattern Under the Plough*, describes how holed flints were routinely suspended above the backs of horses that were 'all of a sweat' in order to attempt a cure.[6] Another more widespread European tradition was to wear them as amulets for protection, or carry them around to promote fertility for both family and livestock. In a similar vein, J. G. Frazer in *The Golden Bough* records them being used in Germany to protect cattle from attack by witches on Walpurgis Night.[7] Hag stones (alternatively written as hagstones) possess as many names as they have potential uses, and have also been known as adder stones, fairy stones and Odin stones in addition to many other local names.

It seems fair to ask what it is about these stones that make them so magical. Is it the hole that holds the significance? Certainly, it is the hole that makes them less than common. Ewart Evans, extrapolating from his folkloric survey of Norfolk hag stones, seems to think that it is because they symbolise the human eye, the hole a substitute for the pupil – a variant of the universal All-seeing Eye that has long been in use for apotropaic purposes worldwide and which is still commonly in use as a painted oculus on the prow of Mediterranean fishing boats and on the walls of Middle Eastern homes.[8] The symbolism is mythical and may hark back to the ancient world: effectively, a hag stone is the Eye of Horus carved by nature. Aleister Crowley, the 'sex magick' occultist and self-proclaimed 'Great Beast' who lived in Hastings in his later years (and claimed to hate it with a vengeance), is said to have put a curse on the town in which those who lived there would not be able to leave unless they could find a hag stone on the beach. For Crowley, the 'wickedest man in the world', who had devised hugely demanding magical rituals to evoke his chosen angels and demons, this seemed a relatively easy way out for anyone wishing to break the spell and take their leave of the generally well-regarded seaside town.

At Cuckmere Haven, not so very far from Hastings as the pebble rolls, the hag stone I had found had a nicely centred, almost perfectly round hole going right through it. The stone was coated with a cream-coloured cortex, and although it seemed a little too large to be carried around all day I could not resist it. Before I put the stone in my rucksack I tried to photograph it at arm's length against the backdrop of the Seven Sisters. Unsuccessfully, as it turned out – a problem with focal length meant that I could not manage to bring both stone and cliffs into sharp focus together. I took plenty of photographs anyway – of the Seven Sisters, of the flints on the beach, of the coastguard cottages on the cliff – the same cottages that frame the classic calendar view of the place.

The beach was busy with day trippers, with Spanish and French tourists as well as local families out for a walk. In the distance on the other side of the river a young British Asian couple was standing apart from the others, hand in hand as they admired the view. They were dwarfed by the vast rise of chalk, the dazzling whiteness of it all. The couple were at some distance away and had their backs to me so I had no idea of what their facial expressions might have been. Even so, they gave the impression of being entranced by the majesty of it all. For all I knew they might have lived locally and visited this place often. Perhaps this was their favourite view, or this was what they thought of when they thought of home? Could it be that they attached some sort of *White Cliffs of Dover* sentimentality to what they saw – not nostalgia or spurious national pride but a visceral sense of territorial belonging and the comfort that home provides. I am speculating wildly, but whatever the truth of it, I found the scene – the framed rectangle seen through my camera lens – profoundly moving. Perhaps it was the scale of it all: the enormous cliffs and the dwarfed human form; the tens of millions of years locked into the cliffs, time-banded by the layers of black flint, juxtaposed against the paltry decades that span even the longest human life. Three million years sequenced in calcium and silicon.

Three million years: roughly the same span of time that stretches from our distant *Australopithecus* ancestors to *Homo sapiens* today. Two spans of deep time that took place aeons apart. In hindsight I can identify my feeling as that same emotion generated on seeing Palaeolithic cave paintings of animals – the gentleness of it, the palpable respect and admiration, the sense of some sort of shamanistic interchange between hunter and prey. How someone living hundreds, even thousands, of generations before, a fellow human engaged in what must have been a precarious hunter-gathering existence, felt the urge to express the sheer joy they felt in being alive in the world. It all amounts to a human connection; across millennia, across deep time. Beauty generates joy. Precariousness spawns religion. Both provoke art in its purest expression.

SEVEN
GAME OF STONES

Grime's Graves – Norfolk | Salisbury – Wiltshire
Belfast, Antrim Coast, Rathlin Island – County Antrim

Flint is nothing if not versatile. With a little dextrous chipping a piece of flint can be fashioned into a tool, a weapon or a sharp-edged blade for butchery. It can also be made to produce something that has value as a beautiful object in its own right. From the Palaeolithic period onwards there is plenty of evidence to suggest that meticulously worked flints were created not solely for practical usage but also specifically for use as votive offerings.

It is difficult to gainsay the way that the Palaeolithic mind thought, or how these people viewed the world they walked in. The best we can do is to make assumptions about some sort of animistic belief system coupled with a shamanistic tradition that attempted to control, or at least work with, nature. Modern anthropology may be able to throw some light here. Did Palaeolithic people, like Australian aboriginal societies, 'sing into being' the landscapes they dwelled in? Storytellers like the Norfolk-based Hugh Lupton have adopted this position to explain the creation of myth and folk tales. Lupton re-imagines west Norfolk's Peddars Way in this spirit, describing the ancient, pre-Roman track, the easternmost strand of the Icknield Way, as a songline that is effectively walked into existence. Certainly, it is a poetically potent way of looking at the world. In Lupton's view, it is a matter of 're-storying' the landscape around us, of imbuing a familiar territory with stories to make it come alive. Following the retreating ice, crossing what was then the land bridge of Doggerland, Lupton speaks of roaming clans that came 'with spears and arrows tipped with shards of flints and carved antler' as they spread out 'across the pale shanks of Albion, a new unfolding into the dream of the land.' [1]

*

In the Palaeolithic and Mesolithic periods, flint was gathered opportunistically wherever it occurred naturally and was usually worked into microliths in situ or at conveniently located sites nearby. With the arrival of farming in the Neolithic, flint started to be mined in a more organised, proto-industrial manner. At the Late Neolithic mine complex at Grime's Graves in Norfolk, miners chose deliberately to dig deep, going down several metres to reach the floorstone that lay beyond the first and second seams of flint. This deeply buried, jet black stone was of the highest quality but, given that the surface stone was perfectly adequate for most purposes, the question remains: why go to the trouble to dig this deep considering the difficulty and potential danger involved? The answer would seem to be that this harder to reach, deeper-mined flint afforded greater cultural value to the objects that were made from it. As previously mentioned, many of the axes knapped from floorstone flint appear to have never been used for practical purposes, and this may never have been the intent as such items were probably always destined for ceremonial use. Such an investment of a scarcer, harder-to-acquire resource, coupled with the skill required to manufacture a high quality axe, made these items valuable exchange goods and the existence of some sort of trade or exchange mechanism is supported by the fact that many have been found at considerable distance from their origin.[2]

As well as finds in mineshafts and graves, Neolithic polished flint axes have been recovered in quantity from bogs and wetlands. There looks to be a strong connection between stone tool votive offerings and wet places both in Britain and in Europe. The Thames is a particularly rich source, although this might be partly down to the attention the river receives from both archaeologists and amateur mudlarks. According to the Museum of London, more than 400 flint axes have been found in the Thames's tidal mud. Although some may have been mislaid simply through lack of care, many were clearly objects of high value that had probably been deliberately placed in the river to demonstrate humility and as an act of thanksgiving to the river deities. The river's confluences, especially, appear to have been favoured locations where votive offerings were presented and worship carried out.

The archaeologist Lars Larsson writes of flint axes being found in almost every bog in southern Sweden: stone tools purposefully deposited after undergoing treatments such as transformation by burning or deliberate destruction by hand. Larsson suggests that such treatment indicates that axe blades held great symbolic importance in the social world of Neolithic communities.[3] He argues that many of the flint tools found were taken out of service in accordance with special rules and norms because the axes had become polluted by use and therefore had to be separated from one social context in order to be integrated into another. The act of deposition may simply have constituted a precondition for increased demand for new material and tools – a sort of market economy perspective in which supply, demand and distribution are kept in equilibrium. The importance and reverence shown towards wetlands in particular may have been because such environments were regarded as liminal zones that connected the physical world with the mystery of the underworld. The presence of water might also be connected with notions of fertility, and so ritual acts such as the deposition of votive offerings would have been made to petition for healthy progeny and encourage the growth of crops. Why some axes were ritually burned by fire is harder to fathom but might relate to the transformation expressed by a subtle change of colour after burning. Flint and fire are of course intrinsically connected and in this topsy-turvy reversal the fire-stone itself is subjected to its own transformative alchemy.

So much of this is conjecture, a donning of the comfortable slippers of the archaeologist's 'ritual use' trope. Many, perhaps most, of us would like to think that these ancient objects held value beyond their practical worth. But what if the makers of these artefacts, incurious to the rhythm of the days, the rise of the sun and moon, the gift of food and water or the curious way that flints flake, just shrugged their powerful shoulders when confronted with the mystery that is life? It seems unlikely, and it is hard to believe that such painstakingly produced stone artefacts were nothing more than just tokens of Stone Age currency.

The best understanding of such belief systems tends to come from anthropology rather than archaeology. Studies of contemporary 'Stone Age' societies, such as those that still exist in West Papua, show that in sourcing jade stone for axe making, inaccessible, dangerous and socially restricted mountain regions are preferred; places that are considered to be liminal and which sit between the worlds of man and the gods. Stone sourced at such places can be considered to be imbued with the power of the mountain it came from, which in turn makes the object made from such stones more precious in the context of trade or ceremonial exchange. In the same way, the polished flints from Grime's Graves, Cissbury and the like that were sourced deep underground could be thought to hold some of that mysterious underworld's power and mystery.

In the Papua New Guinea highlands, Wahgi stone axe makers have been shown to be selective in their choice of site despite an abundance of suitable sources. The anthropologist John Burton explains a threefold balance of factors at play: the geological properties of the stone, the ideology of the axe makers and what he describes as 'the ceremonial competitiveness of the economy in which the stones are exchanged.'[4]

Although polished stone axes had their place in ritual it appears that, for practical purposes at least, flint tools were gradually replaced by their metallic equivalents as the Bronze Age progressed. Evidence from the site of the wooden henge, later to become known as Seahenge, discovered on a beach near Holme-next-the-Sea on the northwest Norfolk coast in 1999 supports this. The site and its remains were painstakingly documented and thoroughly analysed to the extent that using tree-ring data it was possible to date the site to the exact year and the approximate season it was constructed. Built in the spring-summer of 2054 BCE during the early Bronze Age, the site has revealed that a total of fifty-nine bronze axes of different types were employed in its construction. Flint tools of any variety are notable by their absence, which suggests a deliberate exclusion as bronze tools were relatively uncommon at this time, having only been introduced to Britain a few centuries earlier.[5] The employment of bronze tools for shaping wood was cutting-edge technology, quite literally. Seahenge, an oval enclosure constructed of fifty-five closely placed oak posts containing a central 'altar' of an upturned 2.5-tonne bole of an oak, was almost certainly ritualistic in purpose. Perhaps for the builders of this striking and sophisticated monument, flint was thought to be too quotidian, a component of an obsolete technology that was not worthy of an enterprise considered to be sacred? The fact that this community was so well provided with metal axes despite flint being such an abundant local resource makes this even more enigmatic. David Miles in his book on the Neolithic Revolution in Britain, *The Tale of the Axe*, goes even further, suggesting that, 'Seahenge represents the death of the stone axe as a serious tool'.[6]

The use of metal in the British Isles seems to have caught on quickly. While Seahenge provides some of the earliest evidence for timbers worked by metal in Britain, in Ireland they had been in use even earlier, with evidence that the builders of a trackway across a bog at Corlea, County Longford using metal tools around 2268–2250 BCE.[7]

*

Although flint eventually gave way to copper and bronze as a material for practical tools, the polished stone tradition of the Neolithic continued into the Bronze Age, with polished axes often being made using more exotic material like jadeite. Like the deeply mined flint at Grime's Graves, the belief was seemingly the more inaccessible the source, the more valuable the stone.

Such axes were not only valued aesthetically but were of sufficient worth to be traded as either rough-outs or finished objects over surprisingly long distances. A beautifully polished teardrop-shaped axe in Salisbury Museum, dating from the Early Neolithic and believed to have been found at a barrow close to Stonehenge, has been proved to have its origin in a quarry in the Italian Alps.[8] Similar axes of Italian origin have been found in Kent and at Wroot on the (appropriately named) Isle of Axholme in north Lincolnshire. Yet another axe from the same period, found on the island of Jersey, had its origins in the Austrian Alps.[9] Further evidence suggests that, while many were produced on this the largest of the Channel Islands, around one in three polished axes found on Jersey came from the British mainland. Jersey-made axes have been found in England too, which points towards a two-way trade. Neolithic and early Bronze Age people, despite any newfound ties to the land they cultivated, were unquestionably folk who travelled.

The well-documented case of the so-called 'Amesbury Archer' is a case in point. The archer in question was a middle aged man whose grave was discovered in Wiltshire close to Stonehenge in 2002. The 'archer' moniker refers to the large number of arrowheads that were found buried alongside him. Radiocarbon dating points towards his burial being around 2300 BCE, roughly contemporaneous with the erection of the bluestones at Stonehenge. Along with some typical Beaker culture pottery, copper knives and flint barbed arrowheads, the grave also contained material for a strike-a-light, a kit for flint knapping and another for metal working. Cushion stones that might have been used as a sort of portable anvil suggest that the so-called Archer may have been some sort of itinerant coppersmith (and likely flint tool maker too – he appears to have been a pioneer who transcended technological boundaries). Among the largest number of grave goods ever seen at an early Bronze Age burial in Britain were also the first gold objects to be found in England. Most revealingly of all, analysis of the Archer's tooth enamel proved that he was born in a mountainous region of central Europe, most probably the Alps.

Like polished axes and pottery, ideas and skills were commodities that also sometimes travelled far. The Archer, perhaps attracted to southern England by knowledge of the northern cult that was centred on Stonehenge, was not the only well-travelled migrant. It seems unlikely that he travelled alone but rather as a member of a small elite that possessed metallurgical skills and who roamed far with their arcane knowledge and significant wealth. The early Bronze Age people that we now refer to as Beaker folk, of whom the Amesbury Archer was an early representative, had more to offer than just fancy funerary pots. Even so, despite their familiarity with metals, many Beaker graves have been found to contain beautifully worked flint daggers fashioned to imitate the cast bronze tools they forged and used.[10]

While flint was still an important resource for arrowheads and for simple tools like scrapers, the use of the stone now extended to the purely aesthetic: flint worked as simulacra of the bronze blades that had taken over practical usage. With the arrival of the Bronze Age, flint had become ornament.

The Beaker man who achieved posthumous fame as the Amesbury Archer was buried with both stone tools and the accoutrements of early metal working. Straddling a prehistoric cultural boundary, he was an envoy of the new age to come. Archaeologists have named this transitional period the Chalcolithic or Copper Age to differentiate it from the Neolithic or Bronze Age that came before and after. The world's first recorded evidence of copper smelting comes from a site at Rudnik in Serbia that has been dated to around 5000 BCE. In Britain, the Chalcolithic lasted for a relatively short period between 2,500–2,200 BCE.

The bones and grave goods of the Amesbury Archer now reside in Salisbury Museum, a few miles south of his original resting place near Stonehenge. Salisbury is well known for having the tallest cathedral spire in Britain, which at 123 metres is 27 metres higher than its nearest rival, Norwich. Salisbury's town museum, which lies close to the cathedral, was covered in scaffolding when I visited. Some of its galleries were closed for refurbishment, but thankfully the Wessex Gallery, where all the local prehistoric treasures were displayed, was still open. I found the Amesbury Archer on display in a large glass case – *Amesbury Archer, Late Neolithic, 2400–2200 BCE* – with labels pointing at the impressive range of grave goods that were buried with him. These included beaker pots and archers' wrist guards along with copper knives and golden hair ornaments. The display was arranged in much the same way as the grave had been found – the skeleton placed delicately in repose in a curled position. I had been examining the case for less than a minute when I was approached by an attendant who was keen to tell me what she knew about the Archer. She pointed out an absent left knee cap and signs in the jaw bone that indicated that the man had suffered from an abscess. The missing patella would have meant that he walked with a pronounced limp, while the abscess would have caused serious infection and considerable pain, all of which made the arduous journey between the Alps and Stonehenge even more remarkable.

The Archer would probably have been around forty to forty-five years old, a not inconsiderable age in the Late Neolithic. The attendant told me that another grave had been found nearby that contained the remains of a young man probably in his late twenties. This contained a single boar tusk – the Archer's grave had four – and a pair of gold earrings. Whether or not this younger man was a companion of the Archer could not be determined but his grave was more or less contemporary with that of the older man. She confided that the museum staff liked to imagine that the younger man might be the Archer's son who had accompanied his father on the journey across Europe. There was, of course, absolutely no certainty about this although there was evidence that suggested a genetic link between the two men.[11] There was also some suggestion that the Archer may even have travelled to Stonehenge more than once in his life, a hugely impressive feat given the distance involved and the afflictions he suffered from.

What was plain to see was the show of wealth that accompanied the burial, which was probably directly related to the high status the Archer held. This remains the most impressive Beaker burial yet discovered in Britain. Along with the other goods were a total of 122 flint tools and no less than eighteen tanged flint arrowheads, which would more than likely have been originally attached to wooden shafts, although these would have rotted in the grave leaving no trace. The arrows may have been placed around the body during burial, or possibly thrown in by mourners as tribute. The fact that the arrowheads had been unearthed pointing in all directions suggested that the latter may have been the case. I was interested in the origin of the flint but the attendant did not know, although it did look as if many of the arrowheads were made by the same person. There was certainly no shortage of flint locally, or flint tools – an enormous horde of flint tools has been found a site at Blick Mead close to both Amesbury and Stonehenge but these are all of an earlier Mesolithic rather than a Bronze Age date.

I wondered whether the Archer had knapped the arrowheads himself or if they were the work of a third party. It seems reasonable to assume that someone with the ability to work metal would also have been able to knap flint skilfully. Perhaps many men – and indeed, women – would have possessed this skill at the time, so fundamental was it for people for whom hunting was an important part of their existence. On the other hand, skilfully made arrowheads, like polished axes, would have made ideal goods for exchange or barter, and specialist artisans might have done regular business with those who were proficient at hunting but not flint knapping. If this sort of mutually beneficial trade was already in place by the early Bronze Age, then both parties would have desirable goods to offer. The origin of the flint tools may have been unclear, but this was not the case with the copper and gold that was buried with the Archer. Like the man himself, these were of European origin and came from France or Spain respectively.

*

From the Neolithic to the early Bronze Age, worked stone axes travelled far. Continental jadeite axes reached as far as Scotland, where polished Italian axe heads were brought by farming groups from Brittany after already having changed hands many times en route. The series of exchanges that brought the axes to Britain from the Italian Alps spanned a period of several generations, perhaps a hundred years or more. Although the rough-outs taken from the mountains would have been polished in lower-lying Italian settlements before travelling further, some may have been reshaped and re-polished en route. As ceremonial objects that took many hours to make, ownership of such luxury items conferred status and prestige. Symbolic rather than practical, the polished axes were high-value representatives of the original tools that had transformed the environment by allowing their makers to clear land for the planting of crops and grazing of animals.

During the Neolithic period native sources of jadeite or greenstone were restricted to upland areas of north and west Britain: in North Wales, the Lake District, Northern Ireland and Cornwall. Great Langdale in the Lake District, in particular, became an important centre for mining greenstone and hornstone for polished axes, providing a total of 27% of all the polished stone tools found in Britain. As with Grime's Graves, these were mostly in the form of rough-outs that would be completed elsewhere. Langdale axes have been found at sites throughout Britain, even commonly as far away as Lincolnshire and East Anglia. While some of the axes were clearly never used and kept solely for the status they conferred, others show signs of practical wear. Great Langdale was not only remote but it was also a site that required an arduous climb up the scree slopes of the Pike O'Stickle to reach, a factor that probably would have contributed to the perceived value of the greenstone mined.

Other important stone industry sites were located at Penmaenmawr in North Wales and Tievebulliagh and Rathlin Island in County Antrim, Northern Ireland. Like Langdale, these were upland sites where the igneous rock quarried lent itself to polishing. The stone quarried at Tievebulliagh and Rathlin Island was porcellanite, which, as its name suggests, it is similar in appearance to porcelain. As elsewhere, rough-outs were generally made on site before transportation elsewhere. Polished porcellanite stone axes from Northern Ireland have been found throughout the British Isles, from southern England to the Outer Hebrides to the Irish Republic. Most remarkably, a hoard of nineteen large polished axes was discovered along Malone Road in Belfast, some of which were found inserted into the ground. Too big for practical use, some sort of ceremonial purpose seems to be almost a certainty here. The Malone Hoard now has pride of place in Belfast's Ulster Museum close to the site where the axes were discovered.

*

I visited Belfast's Ulster Museum en route to a rental holiday cottage on Northern Ireland's Antrim coast. The museum is set in leafy grounds in south Belfast, next to the Botanical Gardens and close to King's University. A thoroughly modern museum with an abundance of light and polished pine, the rooms of each floor open out onto a gallery corridor beneath a central roof space where *Game of Thrones* monsters dangle menacingly from the ceiling. The Malone Hoard stones are easy enough to find in the prehistory gallery. Well-lit and tastefully arranged in a splayed formation, seen up close they are beautiful objects: perfectly smooth and voluptuously curved, a darker shade of charcoal. For good measure, and perhaps because the urge to handle such artefacts is overwhelming, a couple of facsimiles have been placed outside the sealed display case with the legend *Please Touch*. But these are held in place by metal bands; while the caress of a finger is possible, the opportunity to actually lift and experience the polished stones' glorious heft is sadly unachievable.

Leaving Belfast to drive up to the Antrim coast, I was keen to stop at the village of Glenarm, where exposed flint in the chalk cliffs had once provided an ample supply of raw material for Mesolithic tool makers. It was not hard to spot the cliffs, as large fallen chalk boulders lined both sides of the road just before we reached the village sign. We parked in a lay-by to investigate. Fulmars had made their nests in the cliffs and a few of them circled, straight-winged above our heads. Calcareous plants like cranesbills had colonised the places where long-fallen boulders of chalk had had sufficient time to develop the thinnest of soil layers. The flint nodules trapped in the chalk were mostly a pale grey colour while the beach itself was covered with flint and chalk pebbles alongside those of much darker basalt. Without much effort I soon managed to find a belemnite buried in a lump of chalk, and then shortly afterwards another piece where the same species of Cretaceous squid was held in cross-section like a grey eyeball in the chalk.

Glenarm, with its eighteenth-century castle and cosy Georgian terraced streets, was quietly charming. The small quay had a chalk-block warehouse and a marina that had an anti-Brexit sign on its wall declaring: *No Sea Border.* It was all instantly likeable. The friendly tearoom we visited was playing John Coltrane as background music, although not in a self-consciously hip sort of way. Unlike many Northern Ireland settlements, the mood was sufficiently non-sectarian to make it difficult to determine whether Glenarm had a Protestant or Catholic majority. I learned later that it was 54% Protestant but, like its largely Catholic neighbour just up the coast at Carnlough, the village authorities had decided to remove any sectarian flags in a ploy to boost tourism – Glenarm is one of several Antrim locations used in the Fantasy drama television series *Game of Thrones* and the village is happy to welcome visitors of any religious persuasion.

Fantasy drama aside, the Antrim coast is probably best known for its Giant's Causeway, the intriguing black basalt pillars that fan out staircase-like into the sea near the whiskey-distilling village of Bushmills. The Giant's Causeway is undeniably a remarkable and unique natural phenomenon, but in geological terms, it is an attention-grabbing newcomer. Antrim's geology is actually far more complex, with strata of limestone, chalk and basalt all vying for attention in the high rugged cliffs of its coastline. The younger basalt generally overlays the limestone and chalk, but in places it is eroded away or twisted into unconformities. Most of the beaches we visited were strewn with boulders and pebbles of all of these rocks: basalt, chalk and limestone lying side by side – no stone sectarianism here. Flint was relatively common along much of the coast too, either as beach pebbles or as strata in the cliffs where the chalk was exposed.

The cottage we had rented with friends stood high on a hill a couple of miles outside the small town of Ballycastle. The prospect to the north was an uninterrupted panorama of sea, sky and the long, low profile of Rathlin Island, Northern Ireland's only inhabited island. The view depended on the weather, but the L-shaped land mass that hugged the horizon was always visible even if sometimes it was little more than a vague smudge on the water. On sunnier days it was also possible to discern the island's magpie cliffs of basalt and chalk as well as the houses and church of the village that clustered around the harbour. It was on these brighter days that other landforms also took shape in the distance. We could usually make out the southern end of Scotland's Mull of Kintyre peninsula, and sometimes more distant places came into focus: the hills of Islay and the unmistakable double rise of the Paps of Jura beyond.

From our perspective, the island seemed to be at the edge of things. The most northerly point in Northern Ireland, it was the place where the island of Ireland gave way to the Atlantic. A small islet that lay off the coast of a much larger island, Rathlin gave more the impression of an outlier than a stepping stone to anywhere else.

On a day that promised to be calmer and sunnier than most, we made the short journey across on the *Spirit of Rathlin* ferry. At Rathlin harbour, a bus – the 'Puffin Express' – was waiting to take passengers to the RSPB reserve at the seabird cliffs that marked the island's westernmost point. We spent about an hour or so at the cliffs then, replete with the sight of auks and gannets and the ammoniac stench of guano, we took the next available bus back. Heading back to the village, past fields of grazing sheep and cattle, our driver stopped the bus unexpectedly to point out a cave on our left. It was the entrance, he said, to the stone axe mine that had existed on the island in Neolithic times.

The Malone Hoard axes we had seen in the Ulster Museum in Belfast had been identified as being made from a rock called porcellanite, a hard, impure variety of chert so named because of its physical similarity to unglazed porcelain. Unlike flint, porcellanite does not flake easily and has a texture that accepts a fine polish and keen edge. The two known sources of this scarce rock in Northern Ireland are in Antrim. One is at Tievebulliagh, a 402-metre mountain in the Glens of Antrim; the other, here on Rathlin Island. Both sources yielded rock of identical physical and chemical makeup, and so it was impossible to tell which had been used to make the Malone Road axes. For no better reason than my own geographical bias I preferred to think that these precious objects had originated here on Rathlin. We tend to make our own emotional truths. Irrespective of whatever the source might be, I would always associate the Malone Hoard axes with this windblown place at the edge of the Irish world.

Back at Church Bay we spent a little time listening out for corncrakes, whose presence had recently been reported in one of the nearby fields. The birds were elusive, as is their wont, so we went down to the shoreline at Mill Bay to eat sandwiches and watch the waves. The stony beach, like those across the water on the Antrim coast, was covered with a peppercorn scattering of chalk, basalt and flint pebbles. Beyond this, floating upon the gently lapping waters of the bay were sinuous strands of treacle-brown seaweed, through which the glistening dark heads of grey seals bobbed up periodically like inquisitive dogs.

Seaweed, specifically kelp, was once an important industry on the island, as was the capture of cliff-dwelling seabirds like the guillemots and razorbills we had been watching earlier.[12] The roofless chalk-block kelp barn that stood at the end of the beach was evidence of this old way of life, a place where seaweed had once been stored in bulk in preparation for loading onto boats.

to the mainland. These days it tends to be the holiday trade that supplements most islanders' incomes – there are a couple of bars and cafés as well as gift shops in the rows of cottages that cluster close to the quay at Church Bay. Walking back to the village in search of a cup of tea, I picked up a couple of small flints from the beach. Like most I had found on the Antrim coast, they were light grey in tone, the colour of a herring gull's back. I stowed the flints in an inside pocket and thought no more about it. Too small for their weight to be noticed, I would soon forget all about them.

EIGHT
FAITH IN STONE

Aldham – Suffolk | Norwich – Norfolk | Barsham – Suffolk

O, so light a foot will ne'er wear out the everlasting flint.
William Shakespeare *Romeo and Juliet*

Given the inexorable march of technological progress, it was inevitable that flint tools and blades would eventually be replaced by their more versatile metal equivalents. So would the use of flint and other stones for the votive objects offered to facilitate the passage between life and death. Religion would become formalised and monotheistic – in the 'civilised' world at least – and be based on texts whose literal interpretation would provoke an unswayable pattern of adherence. The control of ritual would reside in the hands of a chosen few. It would, in time, be set in stone – quite literally – as worship would take place inside bespoke stone-built buildings. Often these structures were replacements for the flimsier wooden churches that preceded them. In those parts of England where it was abundant, flint would reappear in the fabric of daily life as a major constituent of the walls of the buildings where the new rituals were performed.

*

Before the building of churches, and the slow acceptance of Christianity as the religion of state, came the Romans. Their 400-year-long tenure of these islands at the edge of their empire brought significant change to the lives of the Iron Age people who were already settled here. Roman culture brought hot baths, running water, central heating, mosaic flooring, paved roads, a calendar, Latin, bureaucracy and rabbits, as well as a pantheon of gods and the concept of violent death as a sporting spectacle. Romans famously also constructed straight roads, villas, walls and fortresses, and introduced brick-making. Many Roman bricks still remain in place today, either in walls and military structures or as recycled material co-opted for use in later buildings. Romans also built with stone and, ever practically minded, they used whatever was locally available. In the south and east of England they made good use of the flint that lay in abundance beneath their sandalled feet. This is true of both the buildings they constructed and the roads that they made, although in roads the flint usually lay beneath layers of other material. As Daniel Defoe notes is his description of an exposed section of the Fosse Way, flint was an integral part of the Roman road-building sequence: 'Clay of solid binding quality, then flint-stones, then chalk, then upon the chalk rough ballast or gravel.'[1]

Building with flint comes with limitations. While its unique flaking properties make it an ideal material for creating thin, sharp edges, flint does not readily lend itself to making rectangles and square corners. Used on its own, without elaborate, time-squandering knapping, it is hard to make a straight line using flint. But flint pebbles combined with bands of Roman brick, levelled up with thick applications of mortar between each course, was an altogether different matter.

Burgh Castle, one of the Saxon Shore forts that the Romans erected in the later years of their time in Britain, is one such building. Although partly ruined, with only three of its original walls surviving, its mighty presence continues to stare out over the confluence of the Yare and Waveney rivers close to the Norfolk–Suffolk border. At the time of its construction, the fort would have been even more domineering, overlooking the much larger estuary that existed before the spit of land that later became Great Yarmouth had developed. At Reedham on the other side of the estuary, these days a quiet inland village on the banks of the River Yare, but coastal in Roman times, they also built a lighthouse. Like the fortress at Burgh Castle, this was made of flint and brick. The lighthouse at Reedham is long gone but some of its original material was almost certainly reused in the construction of the village's fifteenth-century Church of St John the Baptist. In contrast, the walls of Burgh Castle – large flint pebbles set in mortar interspersed with bands of slender Roman bricks – remain impressive, despite having borne the brunt of the worst that nearly two millennia of history and weather could throw at them.

After the Romans left Britain in 43 CE there was a return to building solely with wood and thatch, a tradition that lasted almost until the time of the Norman Conquest. It was around this time that Saxon architects started experimenting with flint in those regions where it was plentiful: in East Anglia, especially Norfolk and Suffolk, and across the southern counties of Dorset, Hampshire, Wiltshire, Sussex, Surrey and Kent. Following the 1066 invasion, it was Norman builders who took control of church building and restoration. For grander ecclesiastic and military buildings they favoured a veneer of limestone imported from the Caen region of Normandy. Even so, they also readily made use of flint as a considerably cheaper and locally plentiful resource.

*

It is difficult to imagine an English countryside without its medieval churches. It is perhaps even harder in counties like Norfolk and Suffolk, where such churches seem to be as much a part of the landscape as the hedgerows, fields and winding country lanes. More than just an expression of religious faith, the flint-built churches that lie scattered across the East Anglia countryside are like buttons on a mattress, pinch points that hold the fabric of the landscape together. Churches stand as markers of the medieval matrix of parishes that criss-cross the rural landscape. Superimposed on the patchwork tapestry of agriculture, they provide focus; they make the landscape human. Although they share common architectural features, each church is unique, a diversity that reflects the myriad ways in which human creativity and environment can interrelate. During the early days of church-building in Norfolk and Suffolk one particular type of feature stands out from other forms: a structure almost exclusive to the region that had been in place even before the arrival of the Normans – the flint round tower.

A flint round tower church features in the opening credits of season two of the BBC TV series *Detectorists.* The church stands in the background of what appears to be an idyllic English landscape of rich green pasture and gently folded valley. The small church looks on benignly as the metal-detecting friends Andy and Lance, played by Mackenzie Crook and Toby Jones, go about the business of seeking treasure in the meadow that lies in the shadow of its tower. The light is perfect, like late-bottled summer; the trees glow green and the meadow grass sparkles with fecund promise. All is well in this timeless rural Eden as the two detectorists leave the quotidian world of fork-lifts and grass trimmers behind to indulge in the pursuit of dreams, male-bonding as they do so in the gentlest of ways. It is all very heart-warming, although all is not quite as it may appear.

The world of *Detectorists* takes place in a fictional north Essex, although the actual location of the scene depicted in the opening credits lies a little further north in Suffolk, where the fourteenth-century church of St Mary's at Aldham can be found just to the north of Hadleigh (the Suffolk town of Framlingham doubles for the fictional town of Danebury, where most of the action takes place). Despite its tranquil, quintessentially English setting, the real-life location is not as untroubled as the one it fictionalises: the hilltop church lies close to Aldham Common where the Protestant preacher Dr Roland Taylor, Rector of Hadleigh, was martyred at the stake during the reign of Bloody Mary in 1555.[2]

It might be argued that it would be hard to find a parish anywhere in the country that had not suffered in some terrible way at some stage in history. Medieval churches, even a modest one like this, stand as witness to religious and social upheaval through the ages, and those watchwords of the medieval period, terror and magnificence, can be applied in equal measure to describe the events that span their existence. Such disparity between perceived and actual history is, in many ways, in keeping with the sometimes disquieting geography of the *Detectorists,* in which it is frequently suggested that, in the English countryside, not all is quite as pure and natural as it may appear.[3] The distinctive church, fictionally shifted in this instance, is merely a marker of location: a flint-towered storehouse of history.

Round towers are an East Anglia speciality; a true expression of the local vernacular. With the exception of a few in Germany, nearly all are in the east of England. Of the 186 that survive in complete or semi-ruinous form, 126 are in Norfolk, forty-one in Suffolk, seven in Essex, three in Sussex, two in Cambridgeshire and two in Berkshire. Almost all date from between the eleventh and fourteenth centuries, and a handful pre-date the Norman Conquest.

What nearly all of these round churches have in common is the material they are built from: flint. The stone may be strong, hard and, in some areas, plentiful, but what it is not good for is making perfectly square corners. Flint with its fine granular form is superb for making sharp-edged things, but it does not lend itself to right angles.

Quite why there is such a predominance of round tower churches in Norfolk, and to a lesser extent Suffolk, is unclear. While flint is plentiful in both of these counties, it is also readily available in chalk country of southern England, where churches of this design are a rarity. There are none in Kent or Hampshire and only a few in Sussex. The explanation is probably cultural. In the past, Norfolk, above all other English counties, had strong trading links across the North Sea and the Baltic with countries like Germany and Sweden. Round tower churches were once popular in the German regions of Saxony and Schleswig-Holstein, and the architectural tradition may have originated there before travelling across the North Sea to Norfolk. Whether or not the round tower was an indigenous Anglo-Saxon invention or a design inspired by builders from across the sea in mainland Europe remains uncertain. The majority of round tower churches were built in the period just before and after the Norman invasion. Many of the earlier towers were built as additions to already existing churches.

Some older theories about round tower churches have been discredited in recent years. They were once considered, somewhat fancifully, by some to be stone-lined wells that had been revealed by a drop of land level following a flood. Others believed that round towers were defensive structures built to provide shelter from attack by Viking raiders. Even the estimable Jacquetta Hawkes, writing in *A Land* in the late 1940s, seems to think that English round towers had both ecclesiastic and military functions. Most round towers would not have been fit for purpose even if this were so and would have been an afterthought anyway, as in many cases the tower was a later addition to a pre-existing nave.

In Europe, a defensive role might have been a possibility. In eastern France and Luxembourg such structures have historically been recorded as defensive towers, and in one instance at Heckenransbach in Lorraine, France a thirteenth-century chapel round tower was topped with a room large enough to hold a whole village in case of attack.[4] Even so, the explanation, for English churches at least, appears to be mostly just a matter of ease of building – corners require square-faced quoins; to build with flint in the round is a far easier option.

The principal purpose of any church tower is to house bells to summon worshippers, a tradition that is sometimes linked to the reign of the Saxon king Æthelstan, who is believed in 973 CE to have imposed a law that decreed that landowners must build a bell tower on their land in order to claim status as a thane. Such churches were privately owned in Saxon times and it was through donations and tithes that owners were provided with an income.[5]

Although flint does not readily lend itself to the building of corners, it has been used in this way in some cases. In some East Anglia churches, quoins for walls were fashioned from large pieces of cut flint. Others have flint for jambs and arches, demonstrating that the necessary knapping skills prevailed even after limestone had become the go-to material for such features from the eleventh century onwards.[6]

There is a tendency to proclaim any church with a round tower as being of Saxon origin but this is not necessarily the case, as the practice for building flint cobble round towers persisted well into the post-Conquest period. Early flint churches were built mostly of un-split rubble, using flints of various sizes roughly coursed and held together with a generous application of stiff mortar. Other material, like ferricrete, carrstone or erratics, might also be employed if they were conveniently to hand. Lime mortar required a considerable amount of time to set, and so individual layers had to be left to harden before another course could be added. Building work also had to be discontinued during the winter months because of potential frost damage. Such breaks in construction are clearly visible in the layers of flintwork. Similar to the annual rings of a tree, these layers, which usually occur at around three-metre intervals in the height of the tower, tell the tale of its construction, its seasonal interruptions and productive periods. They can also show differences between the sources of stones chosen and perhaps even any variation in the skill of the builders employed for the work.

Round tower building continued for several centuries in East Anglia despite the Norman predilection for solid square towers and limestone imported either from Normandy or sourced from quarries like that at Barnack near Peterborough. By the late thirteenth century many existing round towers were topped with an octagonal flint belfry, while some new towers were constructed with a round-profiled lower stage that bore an integral octagonal belfry above it.[7]

*

With over one hundred to choose from, Norfolk has more round tower churches than the rest of England – and indeed Europe – combined. To pick those that stand out from the crowd is a difficult, perhaps even pointless, task. Each one has its own qualities, which is as much a matter of setting and atmosphere as it is of architectural design.

In Norwich, there is the octagonal-topped tower of St Benedict's, which stands close to the city's inner ring road at the end of the street bearing its name. This is made exceptional by what is patently absent: any sign of a nave to go with it. The bulk of the church was destroyed in the bombing frenzy of the Baedeker raids of April 1942. The last of the rubble that remained was removed in the early 1950s. All that remains now is a lonely tower, a focal point for a low-rise housing scheme. It is an orphan on a street that can claim no less than five medieval churches. Of the five – all decommissioned, all flint – St Swithin's is now a thriving arts centre; another, St Margaret's, is an occasional craft emporium; two more, both magnificent, are closed but sometimes used as exhibition spaces. The tower of St Benedict's, though, can only serve as a reminder of the horrors of war – a totem close to a former city gate that is now separated from the hurtling ring road traffic by flinty patches of city wall.

Norwich, famously a city with a pub for every day of the year and a medieval church for every week (more accurately, thirty-two that actually survive), once had a total of six round tower churches in or close to its city walls. One is the aforementioned St Benedict's, whose tower is all that remains. Another victim of the same 1942 air raid was St Paul's, which used to stand close to the Church of St James that now sees service as a puppet theatre. Unlike St Benedict's, St Paul's was demolished completely after the war, its absent footprint making space for a roundabout on the city's inner ring road. St Paul's demise was prompted when the parish was combined with that of a neighbour. Subsequently, its stones were sold off and its valuables taken to other churches in the vicinity. The church and tower along with the eponymous marsh are all now lost to time, an absence marked by a blue plaque.[8]

Of those round tower churches that do remain, St Etheldreda on King Street, with its squat octagonal-topped tower, is now used as an artists' studio, while St Mary Coslany, also deconsecrated, is managed by the Norwich Historic Churches Trust. Another St Mary's – St Mary in the Marsh – is, like St Paul's, a long-gone ghost presence that haunts the cathedral precinct.

The last of the six, St Julian's, is another Baedeker raid casualty, a church that holds sufficient historical importance to have warranted complete reconstruction after the war. Tucked away on an alleyway on the steep slope that leads from the scarp of Brancondale down to the River Wensum, the church is as much a shrine as anything, housing the cell of Julian of Norwich, the fourteenth-century mystic anchoress who shunned the material world to write *Revelations of Divine Love*, the first book in English written by a woman. Old photographs show a round tower castellated at the top; the rebuild is reasonably faithful to the original, although the tower is now truncated and reaches only as far as the roofline. The Norman door that leads to Dame Julian's cell and the fourteenth -century font are both gifts from other ruined or redundant churches. Modern rebuild or not, the church's interior exudes a tranquillity that is hard to pin down. Its soft, green-filtered light seems to confirm Dame Julian's axiom that ***all shall be well, and all manner of thing be well.***

Away from Norwich, there are a number of round tower churches that have become familiar sights on my regular routes around Norfolk and Suffolk. Like the one beside the A140 at the end of the long drag through the linear, trunk-road dissected ('Bypass us now!') village of Long Stratton. St Mary's looks to be a peaceful enough place, tucked away behind trees and sheltered from the constant stream of passing traffic, but truth be told I have never been inclined to stop. There is another round-towered St Mary's, a few miles east of Norwich at the village of Surlingham on the River Yare. The church stands central to a bee-hummed graveyard filled with cow parsley and foxgloves, its hexagonally topped rough flint round tower pleasingly framed by the lych gate as you enter its grounds.

More Mary-dedicated round tower churches stand south of the River Yare at Hassingham, Moulton, Haddiscoe, Norton Subcourse and several other quiet south Norfolk villages. Is it merely coincidence that so many of them are dedicated to the Virgin?

*

There is one round-towered church just across the border in Suffolk that holds particular resonance for me. Holy Trinity, in the parish of Barsham, is located down a quiet lane a couple of miles west of the market town of Beccles. The sluggish River Waveney is close by and beyond the far bank are lines of poplars marking Norfolk field boundaries. In this bucolic location, about as deeply rural as it gets in eastern England, it comes as no surprise to find the grave of Adrian Bell, author of a quintessential work on rural

Suffolk, *Corduroy*, in the churchyard. Corduroy is how the gently sloping newly ploughed fields can appear here, in winter at least, although the economic expectation of two grain harvests a year means that the land rarely stands bare for long these days.

I remember the church and the village from the so-called Albion Fairs of the 1970s. The paddock next to Holy Trinity church was the location for Barsham Fair from 1973 until the last one that was held in the blistering summer of 1976. The fair took place on August Bank Holiday weekend, the same weekend that the rains finally broke. It is this last one that I remember, although my memory of that time is a little blurred around the edges.

There is a website documenting the Albion Fairs that has images of the folksy hand-drawn advertising posters along with a photo album of each year.[9] Scrolling through the images I half-expect to see myself captured there in the prime of youth. But would I even recognise that younger self masked by facial hair and camouflaged in denim? The photos, black and white and colour-saturated Kodachrome, seem both old and ageless. Time has passed and lives have come and gone since the blinking of the shutter nearly half a century ago. A few show the flinty church in the background. With its round tower and thatched roof, it makes the perfect backdrop for what was an attempted revival of a medieval crafts fête, although, as at all the Albion fairs, faux medievalism was unashamedly blended with all the psychedelic trimmings of '70s counterculture. The photos are evocative, especially for anyone who took part back then, and you can almost smell a heady mix of wood smoke, patchouli oil and Afghan resin. They provide a shimmering-edged vignette of the time: local TV news features, unkempt extravagance, jester music, dubious dancing, garish colours, a quaintness at odds with the tweed and check of contemporary rural mores. Good people and worthy intentions, but also bad faith and naive politics. The photos capture a vanished culture: out of time youth resurrecting a romanticised image of those who were out of time before them. Pantomime medieval squalor meets New Age materialist rejection; hedonism posing as enlightenment. In fairness, the fairs never did take themselves too seriously and the whole enterprise was underwritten with a strong sense of playfulness. To attend was to participate in a living, slightly anarchic theatre where to suspend your disbelief was not obligatory. But people out of time, by now.

The church itself, a repository of history in a slowly changing landscape, has an air of timelessness. Its flint walls have witnessed far stranger sights than gently stoned youths in mock medieval attire. In these parts as elsewhere, the iconoclastic rifts that accompanied the Reformation, and later, the savagely implemented parliamentary enclosures and rural impoverishment that followed, are etched into their stone memory. As is, the depletion of labour and hope that came a century or so later with the loss of strong and bright young men to futile trench warfare in Flanders. And at other times, other hardships: the graveyard contains a medieval plague pit, and down the road in neighbouring Shipmeadow stands one of the county's largest Victorian workhouses, now converted to apartments. Time marches on, yet, written in stone, it can also sometimes give the impression of standing still.

One thing sticks in my mind about the fair. On the Saturday, a singer entertained the crowd with joyful West African music – feel-good dance music that went well with the early evening sunshine that bathed the paddock in golden light. At the end of his performance, he said something to the effect of 'I'll see you all in church tomorrow', an incongruous statement given that most of the audience would be more easily persuaded to chant Buddhist mantras or Voodoo incantations than attend a Church of England ceremony. He was true to his word. Next morning, the same singer dressed in the same bright African robes could be seen walking up the church drive greeting and mingling with local parishioners: a resplendent peacock among sparrows. Most of the church attendees would not have dreamed of attending the fair, but there was no need: the fair came to them.

The same church barged its way back into my consciousness in 2021 when the Norfolk musician Laura Cannell released *These Feral Lands Volume 1*, a collection of songs that celebrate legends and myths from East Anglia, Shropshire and beyond.[10] One song, titled 'Barsham Light' and sung by comedian Stewart Lee, was about Holy Trinity church at Barsham. It tells of the how at certain times of year golden light falls onto the crucified Christ sculpture on the rood beam, a biannual circumstance of luminosity that lasts for just a few minutes. The light arrives through the east window; low-slung sunlight steered by flint and ashlar to illuminate the subject of the church's worship. The song seems to suggest this occurs at midwinter, but the event actually takes plaçe around the spring and autumn solstices in late afternoon. Even so, it can be elusive: to witness the event, precise timing is crucial and it is important that there is no cloud cover.

Forgotten about for centuries because the rood screen was moved from its original location, the phenomenon was rediscovered by chance by the Reverend John Buchanan in 1990. He thinks the conjunction of light and Christ sculpture was concocted purposefully, as the east window is placed visibly off-centre to direct the light.[11] The event must be as old as the original fourteenth-century window. The reverend debates whether it was created to suggest an act of religious magic, or simply to serve an agricultural purpose and make worshippers aware that this was the time to move cattle to or from the marshes. For illiterate medieval churchgoers, everything about the building – its flint-patterned architecture, the colourfully painted biblical scenes that once decorated the interior walls, the heady numinous atmosphere created by candle and incense – would have invoked a sense of wonder. Magic was surely the intended outcome.

Absent from my warped 1970s lost-youth reverie about this place was another curiosity. It must have been overlooked or unnoticed at the time; now it might be the very reason for visiting. Holy Trinity, Barsham, with its elusive light show, round tower and thatched roof possesses one more thing that makes it quite special, exceptional even. Its east wall is entirely covered by a lozenge-shaped grid of knapped flint and limestone, a pattern that extends seamlessly as tracery across the window itself. To my eyes it resembles the pastry lattice on a fussy fruit tart. Pevsner describes it as 'a unique conception', and unique it certainly is: a feature that probably dates to the early seventeenth century when the chancel roof was renewed.[12] The flint in the lozenge design has been carefully knapped and shaped to fit snugly with the limestone. A technique known as flushwork, this practice represents the pinnacle of what skilled craftsmen were capable of doing with the stone that lay to hand. By the mid-medieval period, flushwork would become the finest expression of the flintknapper's craft.

NINE

FLUSHED WITH PRIDE

Norwich – Norfolk
Eye, Long Melford, Blythburgh, Covehithe, Dunwich – Suffolk

Round tower churches eventually fell out of fashion, but even as they were still being built, church architects were turning their attention to more refined expressions of the potential of flint as a building stone. By the fifteenth century the necessary use of limestone for door jambs, window surrounds and quoins was extended further with the creation of elaborate patterns that combined limestone ashlar with precisely shaped flint to create what has become known as flushwork. More than any other region of the country, it was in East Anglia that the craft of flushwork reached its apogee.

The predominance of the superb flushwork seen in East Anglian churches results from two main factors. Firstly, there was a reliable and almost inexhaustible supply of good quality flint available locally. Secondly, it was a matter of regional prosperity.

Much of the wealth in medieval East Anglia came from the burgeoning wool trade that created fortunes for the landed gentry and some yeoman farmers. Grazing sheep were foisted upon every bit of available land. Sheep meant wool, and wool equated to money for the landowners. But profit for some resulted in ruin for others, in particular the poor and the landless. Environmental devastation was another consequence in some instances too, as severe overgrazing resulted in the creation of a dust bowl in areas like the Brecks and the Suffolk Sandlings, where light sandy soils predominated.

The expression of personal wealth can take many forms, from conspicuous consumption and the accumulation of luxury goods to the altruism of good deeds and charity. In the medieval period most charitable work was steered towards the politically powerful Church, which was firmly entrenched in everyone's life, rich and poor alike. Piety came at a price and those with money were better positioned to buy indulgences and spiritual favour. The finance of church building was viewed as the ultimate expression of directing affluence towards a noble cause. There was also the spiritual angle to consider. In the fundamentalist, fire-and-brimstone world of medieval Christianity, a wealthy sinner might well wish to guarantee him- or herself a place in the next world. Church building was the definitive means of acquiring medieval 'fire insurance'.

Fine flushwork requires quality materials and a knowledgeable skill set. Such craft does not come cheap. It might even be considered that the finer the workmanship, the more charitable the benefactor. Here was an expression of wealth that worked for both ego and soul: architectural beautification by way of a third party – the Church.

As already mentioned, not all flushwork was made for the glory of God. From the fourteenth century onwards, civic buildings like Norwich's Guildhall were made glorious with the application of precisely knapped flint, but it is in churches that the finest examples of flushwork can usually be seen.

The architectural history of flushwork is long and complicated, but it suffices to say that the craft developed around the late thirteenth century and reached its apogee in the fifteenth. The first known example of flushwork is not found in a church per se but in the entrance to a cathedral precinct. Ethelbert Gate, which leads into Norwich Cathedral Close, is the earliest known example in East Anglia; indeed, it's the earliest recorded that can be positively dated anywhere. The pattern of flint and ashlar squares positioned above the arch of the gate and the rose window motif in the parapet above can both be dated to 1316, although these were restored and the design altered slightly in the early nineteenth century.[1]

Flushwork takes many forms. As well as juxtaposing knapped flint with limestone it can also combine flint with a wide variety of other materials like chalk and crag, or with indigenous East Anglian stone like carrstone and septaria. Brick, too, has commonly been used to great effect both in churches and in farm buildings. Flushwork makes use of a broad variety of patterning that includes banding, tracery, panelling and chequerboarding. It also utilises emblems and motifs like trefoils, quatrefoils and cinquefoils. Another common variant is proudwork, in which the freestone stands proud of the wall to frame the knapped flint within.[2]

*

Fine flint flushwork is the hallmark of many East Anglican churches. Some are so familiar that I pass by them almost without thinking. One of these is St Peter Mancroft, a Perpendicular giant whose tower stands aloft in Norwich city centre alongside the marketplace and the modernist glass and brick edifice of The Forum. The church is a towering counterpoint to the small but elegant Guildhall that stands in isolation on the other side of the marketplace and which was built during the same wool-rich period of the city's history. St Peter Mancroft is a building that

I see almost daily, walking into shopping at the marketplace or having coffee at the Forum opposite. Its broad tower looms over the concrete plain beneath it like a conductor grounding the city's free-flowing energy. Even within The Forum, the tower is a benign presence rising beyond the atrium glass, a mighty pinnacle for diners at Pizza Express to contemplate over their garlic butter dough balls. To see it, solid as a rock; to know that it will always be there, come what may, is a comfort of sorts. St Peter Mancroft's confident bulk gives the impression that somehow the city has coalesced around it, although this is misleading. Built in the fifteenth century, St Peter Mancroft came relatively late in the chronology of Norwich's churches and was the last great medieval church to be erected in the city. The Norman castle on the mound to the east, the Norman cathedral whose spire rises beyond the castle to the northeast; even the marketplace, which was moved to its current location shortly after the Conquest, are all considerably older.

It has to be said that St Peter Mancroft really does not have so much flint compared with most of Norwich's other medieval churches. It has some well-executed flushwork at its lower levels but in comparison, say, to the Guildhall, its civic contemporary on the other side of the marketplace, its flint is relatively modest. Even so, it is impressive. In his highly opinionated yet supremely knowledgeable *England's Thousand Best Churches*, Simon Jenkins awards the church a four-star rating, one of only three awarded in the county (although Walpole St Peter, the 'Queen of the Marshlands' gets five stars). Describing it as 'English Perpendicular at its most mature', Jenkins draws attention to the church's embellished tower, 'one of the most enriched in England, its walls covered in panelling and blind arcading and with three huge three-light bell openings'. He describes a 'superb stone frieze rising to pointed battlements', but does not mention the flint or limestone used to make it.[3] Perhaps that is his point – the supremacy of form over substance? The vast quantity of distantly sourced limestone and expensive knapped flint used to create the flushwork is merely the means to an end. The massive tower with the processional way running through its base is a symbol of power and wealth as much as anything. Built four hundred years after the city's castle and cathedral were built, as close in time to the Industrial Revolution as the Norman invasion, St Peter Mancroft represents a confident nod to the future. Another Simon, Simon Knott, the force behind the commendable *Churches of Norfolk* website, describes it better, placing the church in the context of the city centre it stands in. 'Its massive tower and clerestory like eternal truths rising above the deckchair jollity of the stall canopies, is a constant presence. You can never ignore it.' [4]

The other Simon – Jenkins, that is – makes more mention of flint when describing Norfolk's neighbouring county, although he seems doubtful as to its use as an appropriate material for church building. On Suffolk churches in general he proclaims, '...the rebuilding was splendid and only the materials were poor', adding that 'even the rich patrons of Lavenham and Long Melford had to make do with flint, either set into the walls whole or knapped to produce a flat surface.' He then relents somewhat in admitting that in the execution of flushwork, 'Suffolk is supreme, best seen at Southwold and on the superb tower at Eye'.[5]

The superb flushwork on the tower of the Church of St Peter and St Paul at Eye is something I have witnessed on several occasions. Appreciation of the tower's ornamental finesse is helped enormously by the central location of the church in the town, with the perfect view available from the castle mound directly opposite. The magpie timbering of the neighbouring guildhall is another plus, providing a pleasing juxtaposition of black-painted wood and white plaster next to dark flint and pale limestone. Seen in isolation, the tower is a beauty in its own right, yet it is its location that holds the key, the fields and hedges of the Waveney Valley that slope away gently beyond the town providing the

ideal setting for a perfect union of building and landscape.

Staying in Suffolk, it is the cathedral-like late-Perpendicular church of Holy Trinity at Long Melford that receives Jenkins' only five-star rating for the county. Long Melford, as Jenkins acknowledges, was once one of the richest towns in Europe thanks to the profitable wool trade. A number of the town's cloth tycoons contributed generously to the extensive rebuild at the end of the fifteenth century. Seemingly each tried to outdo the others in the funds they gifted.

The sheer size of the church is also worthy of mention. Not only is Holy Trinity the longest church in Suffolk, but even the attached Lady Chapel is bigger than many parish churches in the county. Holy Trinity's riches extend to superb windows and a number of private chapels, but it is the extensive flushwork panelling that richly embellishes the windows, the tower and the three gables of the Lady Chapel addition that is the most remarkable. The intricate and finely wrought detail extends even to the clerestory, where there is a continuous frieze of text picked out in flushwork. There is also a cornice frieze of carvings and flushwork shields on the battlements. Such detailed work requires individual carving of great skill and accuracy, and the cost of production would have been immense. The transportation of building stone was prohibitively expensive, and it has been estimated that even for as short a distance as twelve miles the price of carriage alone would exceed the cost of the stone.[6] In contrast, the flint used in conjunction with the limestone would have been relatively local and far less expensive to provide.

*

In the space of a few centuries flint as building material had grown in status from cheap and convenient bulk material to a desirable stone in its own right. It could no longer be considered second best to limestone; rather, it had become the perfect complement to it.

In the northeast of Suffolk is another Holy Trinity, less ostentatious than its namesake in Long Melford but equally memorable. This is the church that appears almost magically at the same instant as the reeds and mudflats of the Blyth estuary come into view on the drive between Beccles and Southwold. The church, which stands a little apart from Blythburgh village at the widening of the River Blyth estuary, is a miraculously quiet place given the proximity of the A12, a major trunk road that channels traffic southwards towards Ipswich. Holy Trinity, sometimes affectionately dubbed the 'Queen of the Marshes', is best known for its Seven Deadly Sins bench-ends and the beautifully carved angels that line its rafters and gaze down benignly on the congregation. Some of the angels are said to be riddled with lead shot, wounds carried from the time of the Civil War when local puritan William 'Smasher' Dowsing rampaged through Suffolk on an iconoclastic binge of destruction. Another life-size carved angel, in this case a heavenly village sign, stands outside the church.

Holy Trinity dates from the fifteenth century, but its antecedents are far older, as Blythburgh is recorded as one of the oldest Christian sites in the east of England. An earlier church is believed to have stood on the same spot when, in 654AD, the bodies of Anna (aka Onna, the nephew of Raedwald) the King of East Anglia and his son Jurmin were brought here for burial after their defeat at battle with Penda, the pagan King of Mercia.

Unsurprisingly, this being coastal Suffolk, there is flint galore and the church has some fine panelling and chequerboard flushwork on its buttresses. More intriguingly, the walls bear a series of flushwork letters in Lombardic script, which according to Simon Knott's *Suffolk Churches* website probably spell out a Latin dedication to the fifteenth-century rebuild. One of the most noticeable things about the flint in the church, which is knapped yet quite roughly coursed in places, is the sheer variety of colours on show. Light grey, dark grey, yellowish brown – you wonder over how wide an area the variously coloured flints were gathered as it is hard to believe that this represents nothing more than a random selection of locally sourced stone.

Blythburgh's Holy Trinity is but one of a string of quite exceptionally large, fifteenth-century flint churches stretched along this north Suffolk coastline. Each one displays impressive flushwork. An hour's walk to the east in the coastal village of Walberswick is another, St Andrew's. Immediately north of here, across the river in the well-heeled resort of Southwold, is St Edmund's. Further north still, halfway to Lowestoft, is another St Andrew's at the tiny village of Covehithe. There was once a fifth church at Easton Bavents, but this has long been lost to the ever-encroaching North Sea. Although Southwold's St Edmund's is probably the most noteworthy in terms of its fine architectural detail, the most remarkable to see up close is perhaps the church at Covehithe. Here the bare walls of a once enormous fifteenth-century church enclose a much smaller seventeenth-century one. This is not, as may first be thought, the result of any environmental disaster, but simply a matter of downsizing (a similar fate, and for similar reasons, also befell St Andrew's at Walberswick). Following the Reformation, the parishioners of Covehithe could no longer support such a vast structure and the original church was purposefully derelicted by villagers, who recycled many of the materials to build the smaller thatched church within the walls of the former one. While the original walls are now little more than a façade of Perpendicular arched window above a parapet of chequerboard flushwork, the tower stands complete, dwarfing the smaller, newer church that stands beneath.

The sea is close – too close. This is evidenced by a short walk east along the same narrow road that leads from the coast road to the church. In just a few hundred meters the road comes to an abrupt end as it reaches a cliff. The barley fields that stretch away beside the road also terminate in a similarly abrupt manner. The cliff edge – ever shifting – is as fluid as the sea that encroaches upon it. The beach below, a magnificently lonely landscape of low cliffs, sand and shingle, is littered here and there with the bleached trunks of dead trees. It is a place usually spurned by bathers but popular with dog walkers and flint fanciers – prehistoric hand axes, sporadically released from millennia-long rest in the cliffs, turn up with some regularity. Even in high summer, it is quiet in comparison to the far more popular beach at Southwold, just a few miles to the south, where visitors seem drawn to the proximity of the pier by some unseen magnetic force.

Sometime soon – a matter of years or decades rather than centuries – Covehithe's collapsing cliffs will reach the walls of St Andrew's, and its flushwork will return to the sea, from where it will be carried southwards along the coast by longshore drift. Some of it will join the flints that already make up the shingle beach at Sizewell nuclear power between Southwold and Aldeburgh. Other fragments may show up along the coast at the spit at Orford Ness, perhaps mingling with sea-shaved concrete lumps of red-painted lighthouse. As ever at the coast, there is a state of flux: erosion and deposition, drift and tide; sand, shingle and flint.

Covehithe's St Andrew's will not be the first along this coast to suffer the indignity of surrender to the waves. South of Southwold, the once-thriving and regionally important port of

Dunwich lost virtually all of its churches to the sea in the medieval period. Myths abound that you can still hear ghostly bells tolling offshore on certain tides – apocryphal no doubt, but the sort of thing that you almost want to believe in as they blend with the M.R. James territory of crumbling, ghost-storied coastlines. Mythology sometimes outclasses the unembellished reality of things. And myth is important: a canvas for history to paint its own self portrait. Even so, rubble, masonry and hand-knapped flint all go the same way in the end – a return to the source, from the crafting hand of man to the incontestable embrace of nature.

Today just a small village, Dunwich was once a large coastal township. Hard to imagine now, but in its heyday it was an important seafaring and trading port with many churches and priories to its name. Prosperous in the early medieval period, it had been the capital of the Kingdom of East Angles in Anglo-Saxon times and for a while was even the seat of the East Anglia bishopric. This was all before great storm surges in 1286 and 1287 wreaked havoc by destroying buildings and blocking the estuary that gave the town access to the North Sea. In constant decline since this calamity, of the nine churches that once stood here, only the ruins of a thirteenth-century Franciscan priory remain. The ghostly bells of Dunwich, were they to exist, might bring to mind the eldritch hymn that *SS Titanic*'s orchestra played as the ship slipped tragically beneath the waves.

Whatever became of the flint walls of Dunwich's churches that were lost to the tide, polished by waves, shifted along the shore by drift? Orford Ness, a place that has always held its secrets close, must certainly have fragments locked away in the lower depths of its ten-mile-long spit. The Suffolk coast is, after all, a zone renowned for its disappearances.

*

I paid a visit to Dunwich out of season on a bright late November day. As expected, I found the beach café to be closed and the car park almost empty. In the warmer months the village is usually busy with day visitors but like most coastal honey pots it takes on a different character in winter. The only people to be seen on the sloping pebble beach were fishermen, who had spaced themselves apart from each other in some sort of unspoken mutual agreement, their taut lines framing the giant white golf ball of Sizewell B nuclear power station that loomed on the shoreline to the south. Looking north, Southwold's lighthouse was clearly visible, a white beacon that stood above the pantile roofs of houses close to the pier.

I was standing on the beach, yet the spot that I occupied would once have been in the central part of the town. Spool back 700 years and I would have seen houses and streets where the North Sea now lapped. And churches: St John the Baptist, St Leonard, St Bartholomew, St Martin, St Nicholas and St Peter, alongside what must have been a busy marketplace – all gone now, lost to the ever-advancing tide.[7] All Saints, the last of Dunwich's churches to collapse into the sea, had survived until recently enough for there to be photographs of the ruined building. Years before its final plunge to the beach in 1904, the church had fallen into dereliction as a result of Dunwich's overall decline as a fishing port and gradual abandonment by its population.[8] The tower remained until 1919, when this fell too, leaving just a single buttress standing. A new parish church, St James, was built at a safe distance from the shore in 1832, and in 1924 the remaining buttress of All Saints was transported here, block by block, to be re-erected in the churchyard.

I left the beach to walk inland along the village's single street, passing the Ship Inn, a small museum and a few flint pebble cottages to reach the junction where St James stood. The church takes its dedication from the former leper hospital whose grounds the church stands in. The last of the lepers who stayed here died in 1536. After that it became a hospital for more general ills. It was eventually abandoned in 1685, a century and a half before the church of the same dedication was built to keep its ruin company. All that remains of the leper hospital now are three walls of its twelfth-century chapel. The ruined walls with their Romanesque window arches are mostly of flint and limestone, although large chunks of oxblood-red carrstone and oddities like ship ballast and recycled stone from other ruins have been used for repairs

The reconstructed buttress from All Saints stands alone in a corner of the graveyard. Even orphaned – a keepsake snatched from the eager jaws of the sea – it is an impressive piece of architectural reclaim. With diagonal flint flushwork at its base and vertical chequerboard patterning of flint and limestone blocks, it stands like a dislocated chimney in the churchyard's dappled shade. Bold nineteenth-century graffiti has been carved into the limestone: late visitors to the original clifftop site who wished for their names to be memorialised. At the base is an inscription:

The North West Buttress Of All Saints Church
The Last Trace Of The Churches Of The First East Anglian Capital And See

TEN
HOLDING FIRE

Brandon – Suffolk | Beer – Devon

An emerald is as green as grass,
A ruby red as blood;
A sapphire shines as blue as heaven;
A flint lies in the mud.

A diamond is a brilliant stone,
To catch the world's desire;
An opal holds a fiery spark;
But a flint holds fire.

Christina Rossetti *Flint* 1872

When Keith Flint of The Prodigy sang of arson as he leaped frenziedly around music festival stages back in the 1990s he may have been giving a nod to nominative determinism. 'Firestarter' he was, in name at least.

Flint – the stone rather than the singer – has been associated with fire since time immemorial. Somewhere in the mistiest canyons of the deep past a discovery was made: the same grain-free stone that could be fractured to provide a razor-sharp edge could also create a spark if struck with the right sort of stone. The ability to create fire – a discovery made in parallel across the sparsely occupied hominin world that existed between one and two million years ago – resulted in these canyons of the past advancing to a smoke-filled future. Heat and light were attainable, controllable; the world was alight. Flint was the gift that kept on giving, and fire was perhaps the most precious of all gifts. A miracle for those initiated into its secrets, flint was the most singularly useful stone available throughout the Palaeolithic world.

Although the controlled use of fire was clearly used by *Homo erectus* as long as one million years ago – or perhaps even earlier, as some archaeologists claim – it is uncertain when flints were first used to create fire. The earliest evidence of the use of fire in Britain comes, like so much archaeology that is flint-based, from the Breckland region of East Anglia. Beeches Pit, a 400,000-year-old Lower Palaeolithic site a few miles north of Bury St Edmunds in Suffolk, has provided evidence for the early human (probably *Homo heidelbergensis*) use of fire, alongside a number of Acheulean-type hand axes and assorted knapping debris.[1] The pit evidence here seems to suggest that the fires were kept burning for long periods of time, which might indicate that those who lived here had no independent means of creating fire. Alternatively, it might merely point to a desire for continuity and the creation of a shared environment that was dependably safe and warm – a 'campfire', in the broadest, convivial sense of the word. Certainly, fuel supply was not an issue, as wood for burning would have been plentiful.

Portable so-called 'strike-a-light' stones have been recorded since Middle Palaeolithic times. In Europe, the technology dates back at least 50,000 years, as numerous Neanderthal hand axes have been found in France that have wear marks pointing to their use for fire-making by being struck by another flint or iron pyrite rock.[2] The expression 'give us a spark' remains the vernacular for providing a light for a cigarette in parts of northern England.

For Palaeolithic hunter-gatherers, fire-making was a skill so vital that it could mean the difference between life and death. Providing light and protection, and as a means of cooking otherwise indigestible food, fire was central to everyday life. The lighted hearth, then as now, represents stability, safety, family and companionship. To have the Promethean freedom to create fire at will afforded a freedom, flexibility and a degree of protection from predators that raised humans above other primates. In essence, it was flint's fire-making ability, along with tool-making and the eventual domestication of animals and wild grasses, that made settled human society possible.

With the advent of metal technology, flints continued to be used in conjunction with iron pyrites as a convenient means of making fire. This continued into Roman times, where flint remained the standard means of obtaining fire. Virgil writes of this as a common daily task in the *Aenid: 'Quaerit pars semina flammae. Abstrusa in venis silicicis'* ('Some strike from clashing flints their fiery seeds').[3] Following the medieval period, when the main use of flint was for building purposes, this innate fire-making property would again come into its own with the invention of the flintlock rifle.

*

The name is a giveaway: 'flintlock'. The flintlock rifle, which first appeared in Europe in the sixteenth century, was an enormous step forward for the means by which an individual could inflict harm on his fellow man. It also came in useful for hunting. The flintlock rifle is characterised by its use of a flint-striking mechanism for igniting the gunpowder that would propel a piece of shot, or a bullet, on its way. An instrument primarily of war, the flintlock rifle was the prototype for the many varieties of increasingly efficient and deadly firearms that would be invented in its wake.

Rifles with a flintlock firing mechanism became increasingly popular throughout Western Europe in the seventeenth century, and by the following century most infantry soldiers would be issued with a flintlock as part of their standard equipment. By this time the possession of a flintlock rifle had also become *de rigueur* for many farmers and landowners too, who used them for purposes of hunting and gamekeeping as well as for deterring trespassers.

The classic Thomas Gainsborough painting *Mr and Mrs Andrews*, which hangs in the National Gallery, depicts the male subject of the painting with a flintlock rifle nestled in the crook of his arm. It is a painting that exudes confidence and positions the subjects as king and queen of all they survey. The sweep of the Suffolk landscape captured in the background – rolling green and prosperous-looking, with fields of sheep and harvested wheat – is clearly all part of their tenure. Both Mr and Mrs Andrews, with a hint of arrogance afforded by their privileged propriety, meet the viewer's gaze with an undisguised sense of entitlement. The flintlock rifle serves as a visual prop, an essential accessory to the land-owning stance, as is the dog at Mr Andrews' feet sniffing at the gun barrel, but there is also a strong sense of 'Keep off my land' about its presence in the portrait. The rifle speaks of power, privilege and the threat of violence against intruders. As John Berger has observed of this painting in his book *Ways of Seeing* (1972), 'Their enjoyment of "uncorrupted and unperverted nature" did not, however, usually include the nature of other men. The sentence of poaching at that time was deportation. If a man stole a potato he risked a public whipping ordered by the magistrate who would be a landowner.'[4]

The flint required to fire the flintlock mechanism was sourced from a number of different locations, but, in Britain at least, certain sources became favoured not for only their quality but also for the quantity they could produce. Different sources fell in and out of favour until, around the time of Napoleonic Wars, military gunflint production shifted more or less wholesale to Brandon in the Suffolk Brecks. This area had been mined for flint since Neolithic times at the nearby site of Grime's Graves, just over the county boundary in Norfolk.

In 1790 a contract was awarded to Philip Hayward of Bury St Edmunds to deliver 100,000 flints 'of the best sort' to the Tower of London arsenal. Encouraged by the Duke of Richard, Master-General of the Ordnance and a cousin of Lord Cadogan, who owned the nearby Santon Downham estate where the best seams of black flint were located, Hayward took the decision to move his operation from Bury to Brandon.

The flint knapped at Brandon came mostly from just south of the town at Lingheath and Session Heath, an area that is now a Forestry Commission plantation adjacent to Brandon Country Park. Although flint knapping continued in other parts of the country on a smaller scale it was at Brandon where the industry soon became most prominent. In 1813, fourteen of Brandon's flint masters were contracted to supply no less than 1,060,000 musket flints per month – work that would employ a total of 160 knappers, miners and pit diggers in the town. Brandon possessed both the material and the necessary skills to work it, although during the boom years Brandon-trained knappers sometimes left the area temporarily to make use of flints available elsewhere, as in the early years of the nineteenth century when a number of Brandon men travelled to nearby Icklingham when a good supply of flint was discovered there.[5]

Thanks to the quality of the pure black floorstone from which they were knapped, Brandon-produced gun-flints quickly achieved a reputation for excellence. Such repute was also partly due to the perfection of a 'blade' or 'platform' technique' in Brandon workshops that produced flints superior to the 'wedge' or 'Old English' technique that had been used in previous centuries in places like

Hampshire.[6] The 'Brandon Black' floorstone mined at Lingheath was of a similar quality to that found at the lower levels of the nearby Neolithic-era Grime's Graves mines.

At Lingheath and Session Heath, pits were dug as a series of ledges connected by toe holes in the chalk. The flint nodules found at the upper level, known as 'topstone', was good enough for building but not of sufficiently high quality to use for gunflints. The layer below this provided better quality 'wallstone' that, carefully selected, could be used but it was the bottom level where the floorstone lay that provided the large pieces of superior tabloid flint preferred for knapping. These pieces could be anything up to a few metres across and were generally around 15 centimetres thick.

The mined flint was manhandled up the shaft to the surface before being taken to the Brandon workshops by horse and cart. Here, the work on the raw flint consisted of three separate operations: quartering, flaking and knapping. The first two were carried out on the knapper's knee, which was covered with a leather knee pad for protection. The final part of the process was performed at a wooden block or tree stump. The gunflints produced were of different sizes depending on the weapon: the largest for muskets, smaller ones for pistols. Providing there was a sufficient supply of good flakes to work from, an experienced knapper was capable of producing up to 300 flints an hour. The flint that remained – the cores from which the flakes had been knapped – was then lightly worked for use as a building material. Some cores were squared to size, others rounded. Those which were not large enough for knapping into 'builders' were called 'random faced' and considered good enough for use in flushwork as they had an exposed face. The lowest quality pieces that could be used for building were those irregular fragments left over from quartering. Naturally enough, these 'rough builders' were the cheapest to buy but also the least capable of providing a decent finish.[7]

Flint knapping was essentially a cottage industry, and Brandon flintknappers tended to work either in sheds in their own gardens or, when the weather permitted, in outdoor yards. Ventilation was an inevitable problem, especially in winter when outdoor work was not viable, and the fine dust produced by constantly hammering the flint invariably resulted in serious health problems like silicosis for many flintknappers. Ill health and early death were common. A headline in the *Thetford & Watton Times* in 1916 describes one such death under the headline: *Death of a Flint Knapper*.

The death of Mr Robert Field of 63 Thetford Road [Brandon] which occurred on the 20th December remove the last of the members of the family who have been engaged in the ancient industry of flint knapping for generations. Mr Field, who was no more than 46 years of age, succumbed to the disease peculiar to this occupation. Off and on he had been ill for two years, and his last illness extended over ten weeks. The funeral took place at the cemetery on Friday, the Curate officiating. Enclosing the remains was a polished elm coffin, the plate bearing the following inscription: 'Robert Field, died December 1915 aged 46 years. Peace, perfect peace.'

I went to Brandon to visit the town's heritage centre to find out more about the industry. It was the Saturday of the late Queen's Platinum Jubilee weekend and in celebration of the royal event a coffee and burger van and a sandpit for children had set up in the square by the town hall. To further amuse the children, a street entertainer had been booked to perform some sort of alternative Punch and Judy show using a stuffed donkey as a prop.

Brandon, like neighbouring Thetford across the county line, is not a prosperous place by East Anglia standards. Like Thetford, it seems to have plenty of glum-looking youths in hoodies hanging around, along with a disproportionate number of older folk on mobility scooters. Many of the Union flags that bedecked the town appeared to be more permanent than you might expect for a one-off royal celebration and some of the red, white and blue bunting already looked as if it was starting to fray from constant usage.

As anticipated, there was plenty of flint on show. The East Anglian author Olive Cook, writing in the 1950s, describes Brandon having 'an untamed look about its flintiness, a scent of heath and pine in its air which excites our more primitive instincts.' She also tells of her excitement being 'quickened by the hundreds of swifts whose screams haunt the town'.[8] The swifts these days can, alas, be counted on the fingers of both hands but the flint buildings still remain, like Flint Hall on School Lane, which was once the property of flintmaster Philip Hayward, and the accompanying small cottages which he built to accommodate his team of flintknappers.[9]

On the whole, the use of flint as a building material in Brandon looked to be more practical than extravagant. Many of the houses had courses of roughly hewed half-flints, as black as squid ink – leftovers, most likely, from the gun-flint industry. Some of the more recent buildings were flint-faced too, like a no-nonsense 'Meat Market' butchers' that had a quarter of its glass show window filled by a Union Jack, courses of neatly squared flint framing the windows of the above-shop accommodation. A British Telecom building that resembled a fire station, with windows on first floor and a central garage door beneath, had similar small-squared flintwork. Further down the street, a row of cottages that had been given over to small businesses ('Every Cloud Vape Hut', 'Shampooches' dog grooming service) were built of the same deepest black flint, doors and windows neatly outlined in pale brick. Somewhat ironically, the Flintknappers pub that stands at the dead centre of the town, at the crossroads where its two principal streets intersect, looked to be constructed almost entirely of brick apart from the panel of squared flint above the entrance that held the pub sign. Rather than referencing the town's once prosperous gun-flint industry, the sign depicted a hand clutching an Acheulian-style hand axe, although the fingers looked a little too well-manicured to be believable as belonging to any sort of Palaeolithic ancestor. As for a depiction of the more recent type of flint knapping the town is known for, there was a rustic wooden statue on the bridge of a man in a cloth cap chipping away at a stone on his lap.

I ventured into the small single-storey building that is the Brandon Heritage Centre. The volunteers behind the desk were busy talking to another visitor so I went to look around by myself. Part of the museum is devoted to the flint industry that once flourished in the town. There were exhibition boards describing the three stages of making a gunflint and the tools required for the job. Victorian-era photographs showed some of the town's knappers seated in front of a mound of flint nodules: Charles 'Champ' Palmer (1867–1907), John 'Jack' Roberts Symonds Carter (1880–1957) and William 'Billy' Carter (1857–1904), along with an unidentified male infant. I wondered if there was still knapping work available for the child when he grew to adulthood. I also noted that both 'Champ' and 'Billy' did not enjoy particularly long lives, casualties perhaps of the lung diseases like 'knapper's rot' that commonly afflicted the profession.

Another photo showed two men with some of the flint mosaics that could be seen on the museum wall. More like skilful apprentice pieces rather than anything of great aesthetic appeal, the mosaics were in the form of trotting horses and geometric patterns:

'*Flint mosaics made by William Carter 1857–1904, Flintmaster of The Albion, London Road, Brandon, given to this centre by five of his descendants.*'

A glass cabinet displayed photographs of the knapping industry along with an invoice from a Liverpool shipping company for

100,000 gunflints ('Best quality, as previously supplied'), 20,000 per month at a price of 24/- per 1,000. The gunflints were destined for West Africa, presumably the former British colonies of Nigeria, Ghana, Sierra Leone and Gambia.

As I was about to leave, one of the more elderly volunteer guides asked me if she could help. I said that I was interested in Brandon's flint knapping history and this prompted an enthusiastic response. I was told that knapping continued in Brandon until the she was a girl in the 1970s (without wishing to be rude, I think she actually meant the '50s). As a child she could recall the rattle made by loose flints as they were brought into town on trailers. Most of the flint, she said, came from Lingwood but some also from a place she called Mayday, by which she might have meant Mayday Farm, a place a kilometre or so south of Lingwood on the Elveden road. Most of the knappers were based along the town's Thetford Road, she told me, and, yes, some did work in the yard behind the Flintknappers pub. It was then that her companion started talking about the rabbit fur business, another of Brandon's former industries whose process of skinning and curing was said to swathe the town with an almighty malodorous stench ('I don't think the gentleman really wants to know about that, love'). Eschewing any further talk about rabbits, my guide went on to point out one of the museum's most proud possessions, the original deeds of the Brandon Flint Company written on scrolled vellum and kept in a wooden box. I was also shown a lengthy list of all the Brandon flintknappers that had lived and worked in the town between the nineteenth century and 1975, when the very last of them hung up their hammers for good.

While the industry had started with the establishment of a government factory in the town in 1686, it was not until the early nineteenth century that trade really took off, when gunflint contracts were awarded for the war against France. There are many in Brandon who still hold an alternative view on the maxim that victory in those Napoleonic wars was largely down to the supremacy of the British Navy under Admiral Nelson. Instead, they believe that Napoleon was ultimately defeated by the superior firepower of the British infantry thanks to the efficacy of Brandon gunflints. 'A Brandon gunflint could work for up to thirty, maybe fifty strikes,' I was told by another of the museum guides. 'The ones that the French had would only last for a few shots before they needed replacing.' Given the time it takes to load and fire a flintlock rifle, if this were the case it was easy to see the military advantage that the British troops had. I could not help but think of the United States Constitution with its zealously defended Second Amendment that protects the rights of American citizens to keep and bear arms. In 1791, when the constitution was written, the 'arms' in question would have been flintlock rifles. Cumbersome to handle, slow to load, difficult to aim, such rifles were a far cry from the automatic weapons that can currently be purchased over the counter in many parts of the present-day, gun-blighted United States.

I later discovered that Brandon gunflints were also sourced by the Ottoman army for use in the Crimean War of 1852–4. While it was one of the first conflicts to make use of modern technology like railways and telegraphy, the Turkish Ottoman infantry were still in possession of flintlock rifles, as were some of the Russian reserve units.[10] Even as more technologically advanced warfare ravaged in Europe, other parts of the world continued to shoot at each other using more traditional, and perhaps more 'sporting', means. Trade dwindled as the twentieth century progressed, although Brandon gunflint was still sought after by flintlock owners in far-flung colonial outposts like West Africa and the Indian subcontinent, as well as by antique gun enthusiasts the world over. The last Brandon flintknappers, lacking both apprentices and any promise of future business, finally ceased production in the 1970s.

Spent Brandon gunflints must now litter the soil far and wide across the world's former battlefields, especially those places that were once coloured red for Empire on the world map. Brandon gunflints, distinguishable by their jet black colour, also turn up from time to time on the muddy shores of the River Thames. They are said to be hard to spot and it is only their trapezoidal shape that gives them away in the glistening, clay-rich mud of the London riverbanks.[11]

*

While Brandon was the most favoured source for gunflint from the early nineteenth century onwards, it was not the only one. From around 1700 to the early eighteenth century, the principal source of gunflints for the British Army was from the mines at Chislehurst, formerly in Kent but now part of southeast London. Gunflints obtained from Hampshire and the Salisbury region of Wiltshire were also widely used during this period.

Porton Down near Salisbury in Wiltshire, which has a rich archaeology that dates from the Lower Palaeolithic period, was once home to Neolithic flint mines at a location called Martin's Clump. The flint embedded in the Upper Chalk here was exploited for the production of gunflints between the mid seventeenth and late eighteenth centuries. Unlike the rich supply available at Brandon in Suffolk, there were no fine slabs of flint to exploit here just irregularly shaped nodules, which meant that the process was more wasteful with less than ten per cent of the nodule being shaped into a gunflint. Much of the waste is reported as having been eventually taken away for use in repairing roads.[12]

Another small but important source of gunflint came from the western end of the Jurassic coast in southern England. A short distance west of Lyme Regis, with its harbour, fossils and ammonite-shaped lampposts, is the border with Devon. A mile beyond the cheerful resort of Seaton, the village of Beer is reached by a narrow road that twists down from the A3052. The village name has nothing to do with brewing, but derives from an Old English word for 'grove'. Beer is best known for its smuggling history and the freestone known as 'Beer stone' that was formerly mined here. Beer stone, an easily workable pale-coloured limestone valued for its ease of carving, has been used extensively as building stone since the time of the Romans. It has been used in hundreds of West Country churches as well as prestigious buildings like Exeter and Winchester cathedrals and can even be seen as far afield as Westminster Abbey and St Paul's Cathedral in London, as well as Norwich Cathedral. The manmade complex of caves where the stone was once quarried lies close to the village and can be visited, but Beer stone is by no means the only rock in town.

Enclosing Beer's small pebble beach on both sides of the cove are exposed chalk cliffs with narrow bands of black flint. The most westerly exposed chalk on the entire southwest England coastline, Beer's cliffs have long provided a ready supply of raw flint in an area where, other than secondary flint in the form of beach pebbles, the stone is rare. The high quality flint here has probably been utilised since Mesolithic times. Like Grime's Graves and the earlier Sussex mines such as those at Cissbury, the stone was mined here in the Neolithic period and exported considerable distances from its source – as far west as Carn Brae in Cornwall. Unlike Grime's Graves or the South Downs mines, it is unlikely that mineshafts were ever dug here: it was easy enough to simply remove it from the exposed cliff face.

The utility of flint petered out with the coming of the Bronze Age and the emphasis slowly shifted to that of the locality's other main stone resource. The demand would eventually return in the mid seventeenth century, when Beer flint became a favoured supply source for the firing mechanisms of the muskets of Cromwell's New Model Army. In later years it was also sometimes used as a building material for the village and surrounding area.

*

Visiting Beer on a day trip from Lyme Regis, the village's flinty character became immediately apparent as soon as we walked down the high street to the seafront. This may well have been part of the celebrated Jurassic coast, but the strong presence of the Cretaceous is also noticeably prevalent. As we passed the village pub, The Dolphin Hotel, a building constructed almost entirely of roughly knapped black flint, it occurred to me that here was an establishment that would look completely at home on the North Norfolk coast. So would the terrace of flint cottages that stood on a steep street facing the shore. Onetime fishermen's cottages, they appeared to be mostly used now as holiday lets – another North Norfolk parallel. I counted a dozen in total: all of identical build, mid Victorian; all flint pebble with neat red brick edging around the doors and windows.

Alongside the concrete walkway that led down to the beach steps we saw a mural that gave a little local history about Beer's onetime gunflint industry: 'The largest single supplier of flints for guns at that time', a supply that 'flowed from Beer for almost 300 years.' Another panel on the mural described the Neolithic flint trade in which 'tools made of good quality Beer flint were traded by our ancestors across the southwest.'

The beach held two competing beach cafés with white plastic tables and serried ranks of deckchairs for hire. A few small fishing boats were anchored high up on the shore. I reached down to pick up a couple of pebbles. Apart from a thin layer of cortex, the flint was as black as any that might be found at Grime's Graves or Brandon. Flanking the beach on both sides were off-white chalk cliffs with evenly spaced bands of dark flint that came down to eye level. To the east, the limestone cliffs and pristine white buildings of Lyme Regis were clearly visible across the bay. Halfway along, Seaton's oxblood red cliffs of Devonian sandstone rose above a beachside caravan site. Grey, red and white: a tricolour flag of stone. From here at Beer, four types of rock – limestone, sandstone, chalk and flint – could be seen from a single viewpoint. Three tracts of deep time, too – Jurassic, Triassic and Cretaceous.

ELEVEN
CHTHONIC BOOM

Norwich – Norfolk | Potosí – Bolivia | Grime's Graves – Norfolk

Where you stand, there dig deep!
Friedrich Nietzsche

So much is spoken of psychogeography these days that it has almost become mainstream. Someday soon, progressively inclined schools may be offering it as a bone fide 'A' Level subject. Psychogeography, most commonly described using the Situationist definition as 'the study of the specific effect of the geographical environment, consciously organised or not, on the emotions and behaviours of individuals,' has traditionally been applied to understanding cities and their spaces, although there are some who would also encourage adopting a similar phenomenological approach to the rural landscape. The Situationist definition is by no means the earliest coinage of the term. In

1905, half a century before Guy Debord and the Situationist International were searching for the beach beneath the paving stones of Paris, the American anthropologist Jesse Walter Fewkes wrote that 'Psychogeography deals with the influence of the geographical environment on the human mind'[1], a definition that might be considered more succinct than the Situationist version. Invariably linked to walking, psychogeography is a discipline carried out on the move. Practitioners – city walkers, flâneurs, doom strollers – plod miles of pavement as they scratch the underbelly of the city hoping to tune into unwritten histories.

Having been bandied about so freely over the last couple of decades, the term has become almost passé in its use for describing the interface of humanity and urban space. As a consequence, some writer-practitioners have adopted alternative terms like deep topography, while others have broadened its meaning to refer to almost any emotional response to the built environment. The whole notion of psychogeography seems to have become a little like Zen: if you think you can define what it is then it probably isn't that thing in the first place. Not so much the sound of one hand clapping, it is the gentle scratching noise produced by an introvert earnestly jotting something down in a notebook (preferably rain-soaked Moleskine). I should know.

An interesting variant is psychogeology, a word coined by geologist and historian Matthew Williams, who has applied it to understanding his own city Norwich in terms of the effect the city's subterranean world has on its inhabitants. As Williams says, 'Norwich is a good place where we could pursue a new specialist subject to be known as "psychogeology" – whereby we would study the specific effects of the subterranean city, consciously organised or not, on the emotions and behaviour of individuals and the development of culture.'[2]

Norwich is indeed a good place for such a study. What is not immediately apparent from above ground is that it has long been a city with a strong subterranean element. Beneath its surface layer of medieval churches, ancient and modern civic buildings, car parks, real ale pubs, barista-trained coffee bars and tattoo parlours lies another, largely hidden world where the geology hinted at above the surface can be seen at its most raw. While Paris has its sewers and catacombs, and London, its disused ghost stations on the Underground network, Norwich has its chalk tunnels. The city's bedrock is riddled with mine workings: tunnels dug for extracting chalk and flint, many of which are associated with defunct pits and lime kilns. Some are well documented, while others appear to have been long forgotten. A few announce themselves from time to time by way of a surprise collapse when the city slips in on itself.

*

As noted by Celia Fiennes on her turn-of-the-eighteenth-century tour, nearly all of Norwich's many medieval churches are built largely of flint. So too, despite external appearances, are the city's two principal Norman buildings, the Castle and Cathedral. Although both are magnificent edifices built to celebrate and consolidate Norman power, the limestone that covers both buildings is no more than a veneer, a sleek outer layer applied to a foundation of flint rubble. In the case of the Castle, it was originally only the upper level of the keep and royal palace that was clad with Jurassic limestone, the lower floors being surfaced with dressed flint. It was not until the early nineteenth century that the whole of the exterior was refaced with the Bath stone that is seen today. Caen limestone was the preferred material of the original masons, no matter that the stone required a substantial amount of time and man-power to shift it from Normandy to Norwich by way of boat across the North Sea and along the River Wensum. In later centuries, the limestone used for East Anglian churches would be sourced closer to hand from Lincolnshire and Cambridgeshire quarries like Barnack, although these would eventually become exhausted because of high demand.

While crudely hewn flint was not seen as a suitable material for bedecking anything as grand as a cathedral or royal palace, its local abundance was welcome as a means of providing a base for the imported limestone to sit upon. These huge structures, along with the city's numerous churches, monasteries, civic buildings and the twelve-gated, three-kilometre-long city walls that came later, required the provision of a huge quantity of flint during the early medieval period.

Daniel Defoe, who visited Norfolk in the early eighteenth century, a quarter of a century after Fiennes, was impressed with what he saw in Great Yarmouth far more than in Norwich. He finds the city strangely empty, and observes:

> *The case is this; the inhabitants being all busy at their manufactures, dwell in their garrets at their looms, and in their combing-shops, so they call them, twisting-mills, and their other work-houses, almost all the works they are employed in being done within doors.*[3]

But even hard-to-please Defoe was impressed by the city walls he witnessed in Norwich:

> *Taking in more ground than the city of London... the walls seem to be paced, as if they expected that the city would in time increase sufficiently to fill them up with buildings.*[4]

Norwich's city walls, like the medieval churches, the Guildhall and the Bridewell that Fiennes had so admired, were also constructed largely of flint. Given the limitations presented by medieval transport, the flint had to be sourced from as nearby as possible. There are suggestions that some of the supply may have been recycled from the Roman town of Venta Icenorum at Caistor St Edmund just south of the city but, even if this were true, there was still have been an enormous quantity of additional flint that would need to have to been obtained locally.[5]

The earliest quarrying was probably relatively indiscriminate and took advantage of the material that lay close to the surface on the hillsides that ran through the city. Both chalk and flint had their value: the flint for use as building stone, the chalk for lime manufacture. Conveniently, both were found alongside one other. Some of the earliest quarrying took place on the hillside opposite Bishop Bridge, close to the Cathedral.[6] The manmade amphitheatre that was left behind when extraction finally came to an end in the nineteenth century was later used to accommodate the city gas works. Similarly, the Nest, the original ground for Norwich City Football Club before it moved to Carrow Road in 1935, would occupy another former quarry a little further south on Rosary Road. Another part of the city where early chalk and flint extraction took place was along the steep slope that lies between the ridge at Ber Street and the River Wensum. This area was covered by terraced housing in the Victorian period, but as a result of the ground instability caused by quarrying it was mostly cleared in the following century.[7] Structural problems, chalk tunnels and city housing would go on to have a close association in the city. In Norwich, building surveyors are rarely short of work.

As Norwich grew in population and became increasingly congested, quarrying within the confines of the city became more problematic. In time, the extraction of chalk and flint was obliged to migrate beyond the city walls – west along the slopes of the Wensum valley upstream of the city, along the valley of the long-vanished River Dalymond that once flowed into Fishergate from north of Norwich, and along the valley of the River Yare that skirts the city to the south.[8] The demand for chalk and flint continued undiminished into the late medieval period. Eventually, as surface quarries became exhausted or threatened to undermine the cliffs they were penetrating, it became necessary to start tunnelling underground.

It was an incident in 1988 that elevated Norwich's hidden

and largely forgotten underworld into the public eye. This was the occasion when a double-decker bus plunged dramatically halfway down a sinkhole that had suddenly opened up on the city's Earlham Road close to the Catholic Cathedral of St John the Baptist. None of the bus passengers were seriously injured, but the visual comedy of the incident – a half-submerged double-decker bus at 45 degrees to the horizontal – did not pass unnoticed. The bizarre event was immortalised in a photograph that was later used in an advertising campaign for a Cadbury's chocolate bar under the banner '*Nothing fills a hole like a Double Decker.*' The sudden appearance of a sinkhole in the city had come as a surprise, but it was really nothing new. Half a century earlier, a 25-metre-deep sinkhole had opened up and swallowed three houses on Merton Road west of the centre, killing two and making three families homeless.[9]

The Earlham Road area close to St John's is particularly rich in underground tunnels, although there are known systems located all around the city. Only partially recorded, the chalk tunnels that lurk beneath the ground in the Earlham Road and Rosary Road areas of the city are roughly mapped thanks to later surveying. In other parts of Norwich their location has been forgotten or merely rumoured as hand-me-down folklore. A map of the Earlham Road tunnels, based on an original 1823 plan, was published in the *Eastern Daily Press* in 1939. The underground tunnels had been given names, some quite grand, like Royal Arch, Great Temple Street and Discovery Street. Whether or not these were simply the invention of the mapmaker is hard to tell. Unfortunately the plan in the newspaper was misleading, drawn at the wrong scale and orientated incorrectly in terms of the modern street layout. Even so, the plan demonstrated the extent of Norwich's tunnel system, one of several complexes beneath the city streets.[10]

*

Chthonic Boom

A decade or so ago, completely by accident, I happened upon a chalk tunnel entrance in Norwich. Straying into an area of woodland south of the city next to a golf course and an area of urban fringe marshland I came by chance upon a steel door that had been exposed in a low cliff face. The vegetation growing around the entrance had been roughly cleared and the door that sealed the tunnel had quite clearly been prised open. The opening led to a tunnel that lay beneath the neighbouring golf course. Whoever had taken a jemmy to the door had left a gap sufficiently large to squeeze through. An enticing glimpse at Norwich's forbidden underworld beckoned. I am not an urban explorer by any stretch of the imagination, but here was an opportunity that was hard to resist.

The first thing to do was to search the internet. Here I discovered a lively community of derring-do Urbex enthusiasts. Some had posted YouTube videos and talked of 'crews' working together (expedition birders use the same term – there are definite parallels to be drawn here). For some reason, like birders and metal detectorists, many chose camouflage clothing for their exploits. The mood was uniformly masculine in tone, generally sombre and occasionally veering towards the dark side. After much googling I came across an account of the tunnel system whose entrance I had found. There was even a rough map of the layout of the systems various passages, of which I made an even rougher sketch. The account came with a slightly sinister warning: the crew had withdrawn from exploring some of the deeper tunnels of the system when one of their members complained of breathlessness and dizziness.

I summoned a friend to come along with me, both for moral support and as an accomplice to keep an eye on the door. Rather than any atavistic fear of the dark it was the thought of the entrance being resealed – highly unlikely, of course – and my being trapped in the tunnel that concerned me most. My friend stayed near the entrance while I entered with some trepidation, holding a torch, a camera and the scrap of paper with the scribbled tunnel map clutched in my hands. Even within a few metres of the entrance the darkness was absolute and intimidating, the light of my torch too feeble to identify much more than the tunnel floor and highlight the spray-painted arrows and crosses on the walls that signed dead ends and potential exit routes. I took a few photos using flash by simply pointing and hoping for the best: a strange turn of events in which the results viewed later revealed detail unseen by dim yellow torchlight at the time of their taking. Not that there was much detail to be seen other than off-white chalk, fragments of flint and the occasional ciphers of those who had been before. I walked a little further in and, only half-trusting the map I had, took a few turns left and right to complete a short circuit that brought me back to the tunnel close to the entrance. It came as considerable relief when I saw the shaft of light that penetrated the gap in the steel door at the entrance. It felt good to have ventured a little way, but by now I had come to the conclusion that this was no place to linger – or worse, to become lost.

*

For those who spend their entire lives above ground – effectively, almost all of us – the subterranean world can bring out deep-seated fears. To venture below ground is to enter an alien and challenging environment, an unnatural place for surface dwellers. The world beneath the surface is more than just a physical location; it is a forbidden place, an 'underworld' with various dark psychological and spiritual associations.

I had been underground before, but these past visits had been to well-lit caverns filled with phantasmagorical stalactites and stalagmites, like those in Derbyshire's Peak District or the karst country of western Slovenia, or tropical caves in south-east Asia populated by strange blind fish, albino insects and bats – beautiful and mysterious, but largely unthreatening spaces. Apart from a well-managed tour to a West Midlands coal mine at the Black Country Living Museum in Tipton, my only other previous experience of a working mine was a visit twenty years earlier to a terrible, yet intriguing, place on the other side of the world.

Cerro Rico ('rich hill') near Potosí in Bolivia is a silver mine that stands at over 4,000 metres altitude in the *altiplano.* I had signed up for an organised tour in the company of half a dozen Australian backpackers. After presenting our gifts to the guide – cola, coca leaf and sticks of dynamite bought from the city market – we were shown around some of the tunnels. As expected, it was a thoroughly oppressive environment – hot, airless, cramped and tangibly dangerous. We scrambled around the mine's claustrophobic tunnels for a while, dodging the waste-filled rail wagons that rattled past at speed, before being taken to see *El Tío*, who turned out to be a badly sculpted clay figure with goat horns, a woollen hat and a cigar in his mouth. It was the briefest of glances, and I got the distinct impression that we were not supposed to linger as 'The Uncle' was something very personal to the miners and not meant for tourist consumption. It felt almost as the mere presence of gringos like us would eat away at *El Tío*'s magic and somehow diminish his power.

*

In South America *El Tío* is the guardian of the underworld, the spirit that must be appeased before entering his realm. If *Pachamama* can be described as the Earth Mother goddess who controls earthquakes and presides over the fertility of the Earth, then *El Tío* is her drunken, good-for-nothing husband. *El Tío* represents a folksier version of the Devil, a horned figure who was not downright evil but who could be malevolent when he felt like it, which tended to be quite often. The Bolivian miners at Cerro Rico are nominally Roman Catholic and, naturally, any representation of *El Tío* is forbidden by the Catholic Church. Below ground though, things are different: the Church has no dominion in this realm and it is cigar-munching, aguardiente-drinking *El Tío* who is believed to reign supreme.

Many Old World mining traditions used to have their own version of this lord of the underworld. It seems as if the old gods – those worshipped and feared by the likes of Romans and Celts – manage to hang on longer underground, continuing to hold power in their subterranean realm long after the surface world has given way to Abrahamic belief. Miners, like sailors, have always had their own strongly held superstitions and unwritten laws that needed to be adhered to, and in the past many European miners held strong beliefs concerning the existence of underground spirits. This is particularly pronounced in British and German folklore. In the British tradition, especially in the Celtic regions of Wales and Cornwall, the most celebrated supernatural being was the Knocker (*Bwca* in Welsh), a short grizzled creature that bears some similarity to the mythical Irish leprechaun. A European equivalent of this was the Kobbold, a commonly accepted belief in sixteenth-century Germany. The name derives from the sound of knocking on walls that warned of an imminent tunnel collapse. Some miners considered the Knocker to be evil while others believed that they were the benign spirits of those who had previously died in mine accidents but had remained in their place of work to warn others of impending disaster. In Cornwall, miners would leave the last bite of their pasties to the Knocker in gratitude for their watchfulness.

Mining has always been a difficult and highly dangerous pursuit, a reality which those who practice it are fully aware, and there is plenty of evidence to suggest that it has always been a profession in which some degree of propitiation is required. Aside from the obvious physical dangers there is a strong psychological element to mining in which the miner is confronted with the unnatural darkness and claustrophobia engendered by entering the subterranean realm. Neolithic flint miners, with their very basic equipment and total lack of safety provisions, were probably not so very different in their outlook to classical Greeks with their chthonic cults; or, for that matter, with modern industrial man and his inherited folklore beliefs.

Modern day miners are by and large poor people who do the bidding of the rich because they have little opportunity to escape the treadmill of poverty they find themselves trapped in. The dichotomy of rich and poor, danger and safety, privation and luxury appears to be almost universal, as is the sense of community and brotherhood engendered by working in such a tough profession. Whether it is a Bolivian silver mine, Black Country coal mine or Norwich chalk pit, the constant discomfort and ever-attendant danger is a factor that is universal, as is the physically exhausting work entailed in the process. Mining is a tough, dangerous profession, and often a combative one. Dogged and proud, too: the acrimonious Yorkshire miners' strikes of the Thatcher years are a relatively recent memory. But, we might ask, has mining always been a low-status profession? Were Neolithic flint miners, those pioneers of the underworld, put-upon individuals on the lowest rung of the hierarchical ladder, or were they part of an elite group who gladly performed a prized task? Or were they simply opportunists who took advantage of the resources available in their immediate environment? Is it right even to impose contemporary values on the sort of status these people might have had?

The working class culture that permeated the coal mining industry of the last century has been explored in fiction, as in *How Green Was My Valley* by Richard Llewellyn, which gives a nostalgic view of a mining community in pre-war South Wales. It has also been covered in more overtly political works like *The Underground Sea* by John Berger, which deals with the United Kingdom miners' strike of 1984–85. A rare attempt to concoct a prehistoric mining scenario can be found in *The Gift of Stones* by Jim Crace, a novel that centres on a fictional Late Neolithic mining settlement near to the coast in southern England.[11]

Crace's flint-working community is hardworking but insular and suspicious, un-inclined or unwilling to explore beyond the village where they relentlessly mine and knap flint, unable to imagine a future that might be different in any way from what they have always known. The knappers' contemporaries are the merchants, who trade but do not labour, and enjoy a higher standard of living than the miners whose work they exploit. The book works as a metaphor for modern capitalism, or more specifically Thatcherite trickle-down, free market economics (the book was published in 1988, two years before Thatcher's resignation). There is even the suggestion that it may have been during the Neolithic period that capitalism, in the broadest sense, came into being. The conclusion of *The Gift of Stones* comes with the arrival of aggressive bowmen bearing bronze-tipped arrows. It is the dawn of the Bronze Age. The trade in flint goes into rapid decline as a result of this meeting of old and new cultures. Inevitably, it is the flintknappers who suffer the most because of the technological redundancy of their goods. In contrast, the merchants, who are quick to invest in this brave new world of smelted metal, thrive.

*

There was one mine I wanted to see again. I had already visited the Neolithic flint mines at Grime's Graves in Norfolk and enjoyed the standard visitor experience: a descent down a steel ladder to visit the single pit that was open to general admission. During that visit I was told that, later in the year, there might be the possibility of a guided visit to the Greenwell Pit, a larger pit that was normally closed to the public.

The Greenwell Pit takes its name from Canon William Greenwell, a Victorian cleric with an interest in archaeology who had excavated the shaft between 1868 and 1870. It was Greenwell who had been the first to successfully identify Grime's Graves as a collection of flint mines from the Neolithic era. The pit was re-excavated in the 1970s by the British Museum, when previously undiscovered galleries were excavated and several new discoveries made. Among these were the bone remains of animals like ox, goat, sheep, horse and pig, which suggested that the miners were eating, or even perhaps feasting, in the pit as it was backfilled. There was also evidence of a hearth. Among debris that included a large scatter of antler picks, a polished greenstone axe was discovered lying on the pit floor. This appeared to have been carefully placed between the tines of red deer antlers. Also positioned between the tines, just as painstakingly sited as the polished axe, was the skull of a small wading bird, a phalarope.

Both the bird skull and axe were exotic in this corner of Norfolk. Phalaropes were not native to East Anglia and were probably scarce in the region even in Neolithic times. The greenstone axe, easier to pin down geographically, had been brought all the way from Cornwall. Groupings of ritualistic objects like these appear to have been put in place at the abandonment of individual pits, evidence perhaps of a closing ceremony and renewal rite. It was the phalarope that intrigued me most. A scarce bird usually found only in Scotland, what did it signify here, dislocated, twelve metres beneath the soil of a landscape that was far removed from its natural peat bog and shoreline habitat?

Several months after my first visit to Grime's Graves, an email arrived inviting me to join a small tour of Greenwell Pit. It involved the payment of a nominal fee and came with the warning that I would need to wear a safety harness to descend the shaft. It was an opportunity I could barely refuse.

I arrived slightly late at the visitor centre on the prescribed date as a result of a tractor tailback en route – a typical Norfolk scenario. I was the last to join a group that comprised of just two other visitors and a guide. Suitably kitted out with helmet and harness we walked across to the pit at a corner of the site, skylark song ringing high above us as curious sheep eyed us suspiciously. Reaching the entrance shaft, we took turns to be hooked to a safety line before descending the ladder to reach the pit bottom.

A number of side galleries led off in various directions at the base of the shaft. Our guide pointed out the age-blackened pit props that supported the low roofs of the galleries. These, he told us, had been in place for around fifty years. The important thing to remember was to *not* touch anything other than the floor. Definitely *not* the walls or the pit props, or any of the antler picks that lay about on the floor. As a general introduction, he spoke about the supercontinent of Pangaea and its gradual breakup over time, before moving on to describing how flint was formed by the slow accretion of silicon and chalk in warm seas 90 million years ago. He then spoke in detail about the history of Grime's Graves, pointing out the rich black layer of floorstone flint that circled the bottom of the shaft, the prime reason for Neolithic miners digging this deep.

It seemed a good moment to ask the question that had been bothering me about the mines. Why here specifically? And why go so deep when a plentiful source could be had with far less effort close to the surface? I was told that the large blocks of floorstone that could be obtained at this depth were of a vastly superior quality for knapping, a fact verified by modern-day flintknappers who described working this finest of flints as like 'cutting through butter'. In contrast, the flints that lay close to the surface were often frost-damaged to the extent that they were more brittle and broke more easily. I was curious as to how Neolithic miners would have known that stone of this quality existed at this depth as there was no obvious evidence on the surface. The explanation was that before drifting sand created a surface layer over much of the Brecks region, pockets of floorstone would be exposed as strata in chalk cliffs and this would have indicated the depth they were required to dig down to reach it.

We talked about the extraction process: of baskets being lowered into the shaft to take the mined flint to the surface and of back-filling galleries with rubble once they became exhausted – a procedure that avoided having to remove waste from the mine.

The Grime's Graves pits were actively mined for about 400 years during the late Neolithic period. Despite the impression of this being a large industrial site, all the evidence suggests that there would only ever have been one or two pits operational at any one time. The workforce would probably have only been around fifteen in total. Rather than a frantic pillaging of the earth, it was a slow, deliberate extraction process that finally came to an end when the demand for flint diminished with the onset of the Bronze Age.

I mentioned the ritual offering of the antlers, greenstone and phalarope skull. This, our guide agreed, was almost certainly propitiation of some sort: an offering made out of gratitude for the flint. He also suggested that it might have served as a demonstration of the geographical range over which the flint would be exported – from Cornwall (the greenstone axe) to Scotland (the phalarope skull). I thought this second point a little fanciful but, as with any archaeological evidence connected with what appears to be ritual use, it is wide open to interpretation. The key thing was that the ritual arrangement utilised objects that were of great value to those who had placed them, a worth determined by both their rarity and the distance they had travelled to reach here.

We crawled along a couple of the galleries that radiated from the central shaft. In one, the pointed end of a small belemnite protruded from the chalk like an arrested bullet – the fossilised guard of an ancient, extinct species of squid locked in time and space. It was difficult to comfortably stand, and so most of our movement was made either crouched down or by shuffling along on our knees. Under instruction from our guide, we switched our torches off for a minute to experience the darkness of the pit. Once our eyes slowly adjusted to the gloom we could just about detect the barest glimmer of light bleeding in from the open central shaft. It was humbling to acknowledge that this would be how Neolithic miners would have carried out most of their work – in near-darkness, for the most part.

After about forty minutes caught up in the adventure of being in such a challenging and alien environment, something very strange happened: a chain of events that unfolded so suddenly and unexpectedly that my immediate reaction was one of mild bewilderment rather than anything more obvious like fear or terror.

We were in one of the galleries close to where it opened onto the central shaft when, without warning, a few lumps of chalky rock fell from the ceiling above. Then, an instant later, more pieces fell as one of the pit props collapsed. The wooden prop landed on my shoulder, not particularly heavily. As this happened, I must have inadvertently moved and struck my head on a piece of flint that protruded from the chalk wall of the gallery. In the same instant, another pit prop collapsed to fall on the leg of another in the group, although he quickly lifted it off without fuss and appeared to be unhurt. Meanwhile, the third member of the group – a well-built middle-aged man – rushed past making for the entrance to the central shaft. I sat dazed for a few seconds before being beckoned by the guide to move to the safety of the central shaft. My safety helmet had been knocked off and I thought at first I had lost my spectacles before I realised that I was still wearing them. The dislodged pit prop had left a black mark on my jacket but had not hit me with any force. My ear was sore, though, and it was then that I realised it was bleeding.

We were quickly evacuated. As the only one with any apparent injury, I was the first to climb the ladder. Despite the blood, my wound turned out to be no more than a small gash at the top of my ear at the place where my spectacles rested (it was the soreness I felt here that may have provoked the original sensation that I had lost my glasses). Back at the reception area we were given tea and biscuits as accident report forms were filled out. While my ear was being cleaned up I had time to reflect on what had just taken place. Some of it was already a blur. What surprised me most was that given the drama and potential danger of the incident I did not experience any great sense of fear.

The mind has surprisingly effective coping methods for dealing with trauma that have nothing to do with bravery. The subjective distortion of the passing of time is an important element of this, and the temporal fabric of the day had been torn. The last two hours had been both squeezed and stretched. Time had gone soft – temporarily – and my recall was already partly selective. It was not until Jackie arrived to collect me later that I started to experience a more tangible sense of shock.

The shoulder of the light-coloured fleece I had been wearing was blackened by the rotting wood of the pit prop (those watching me climb the ladder to the surface were concerned that it was congealed blood – thankfully it was not). Emptying pockets the next day to put the jacket in the washing machine, I removed two small flints I had gathered from the shore at Rathlin Island. These were stones I had picked up a few weeks earlier and forgotten all about: flints that had unwittingly been taken across water to another flinty place – an unintentional transportation of votive objects. Had I remembered that I had the flints about me I might have offered them up in some fashion – a poor substitute for Cornish greenstone or a phalarope skull admittedly. Anyway, it was too late now. I had been bloodied. A flint-cut ear was my penance. Had this been a warning, or an initiation?

TWELVE
NEST OF STONES, NEST OF ARTISTS

Norwich, Happisburgh, Wighton – Norfolk | East London

The first hole made through a piece of stone is a revelation
Henry Moore

Palaeolithic hand axes like the one found at Happisburgh have a timeless grace about them. The question is: should we be surprised that we can see sophistication and beauty in an object that is 850,000 years old? An object made by a more 'primitive' version of ourselves, a long-extinct version of our species. Does its beauty alone make it art? The answer is almost certainly no, but somewhere along the long forked road of human evolution a consciousness developed in which an aesthetic awareness came to life. Mere practical necessity is not sufficient to explain the delicacy of form of some hand axes. Nor the perfect symmetry seen in many examples, an unnecessary refinement for what is effectively a simple scraping or cutting tool.

For prehistoric people, stone – flint – was more than a medium for tool-making; it was what they thought with – their toolkit, their arsenal; their sole means of shaping the world around them. Stone was a necessity for life but also a vehicle for expression. A bifacial hand axe is effectively a portable piece of sculpture as well as a functional object. And sculpture is art.

Sometimes the stone itself provided ready-made aesthetic interest. Pieces of flint with naturally occurring features like holes or fossils were not always rejected as being too difficult to work; sometimes it appears that they were seized upon as a creative opportunity. And sometimes the irregularity was highlighted, integrated into the finished tool. A fine example of this comes from West Tofts in Norfolk, where a flint axe estimated to be between 300,000 and 500,000 years old was found with a fossilised *Spondylus spinosus*, a bivalve mollusc, at its centre. Not only was this probably a deliberate choice on the part of the axe's maker but it is clear to see that a few well-aimed blows have been executed around the fossil to display it more prominently. It is hard to imagine that such an artefact did not carry special meaning and perhaps even a sense of pride for the individual who created it.

As a raw material, flint has often provided the medium for objects that appear to bridge the divide between the purely functional and that which is aesthetically pleasing. Pre-Colombian Mesoamerican cultures like the Maya have left behind a myriad of flint artefacts that appear to belong to a world that lies beyond mere functionality. So-called 'eccentric flints' are remarkable creations that have been identified as ceremonial in nature. These are generally made from flint's closest geological relatives – chert, obsidian and chalcedony – although they are usually simply referred to as 'flints'.

The most obvious thing about these objects is the degree of skill required to make them. Often they take the form of animals like snakes or scorpions, or multiple-headed human figures, their heads emerging from the body on 'tangs'. While some may have been fixed to a wooden staff or handle and carried, the fragile complexity of the form suggests that they were mostly created specifically as offerings. In the purest sense, they might be said to be examples of religious art. The fact that these objects were often found buried under temple altars supports such an interpretation and parallels similar burial offerings made of flint, porcellanite or greenstone during the Neolithic period in the Old World. In the New World, many aspects of what was ostensibly a Neolithic or 'stone age' culture continued until the Spanish Conquest in the late fifteenth century.

Anthropologists tend to associate eccentric flints with the Mayan lightning deity K'awiil. In the same way that prehistoric hand axes that were thought to be thunderbolts in pre-Enlightenment Europe, Mayan culture considered obsidian to be fossilised lightning. Lightning was firmly linked to agricultural fertility, and specifically maize, the staple crop of the Maya and the most fundamental bringer of life.

Flints and related material were also worked in similar ways in Old World societies, albeit for more pragmatic purposes. During the Early Dynastic Period of Ancient Egypt, before the introduction of iron technology in the first millennium BCE, shards of flint were frequently used in the manufacture of sickle blades. Typically, a number of serrated flint pieces would be slotted into a wooden haft and secured with adhesive to provide a tool capable of reaping the wheat grown in the delta and along the course of the Nile's floodplain.[1] As in Mesoamerica, flint or a similar related material played a key part in grain production, although the eccentric flints of the New World appear to be more closely linked to propitiation than to any practical farming purpose. Even so, Egyptian flint sickles have often been found deposited as grave goods, along with loaves of the bread they helped to produce.

*

While flint in its natural form has inspired artistic endeavour as a medium, the stone's aesthetic character can also be shown for what it *is* rather than for what it might become. Norwich Castle Museum has a John Crome painting dating from 1811 that depicts two flint nodules as a still life. It was a bold departure from the classic subject matter of bowls of fruit, and the focus on flint gives the painting a distinctly local feel. This was probably the intent, as Crome, like all members of the Norwich School of painters that he helped found, would have been keen to give the observer a clear idea of a sense of place. After all, nothing quite says 'Norwich' like a lump of flint. The painting, entitled *Study of Flints*, depicts, centre stage, a large broken flint nodule, the stone's white cortex contrasted against its coal-black interior. Alongside it is a smaller, slightly browner piece, along with a couple of common wayside flowers – dandelion and daisy. The undulating landscape in the background looks like that of a sandy heath. It might well be Norwich's own Mousehold Heath, although there is nothing to confirm this other than the knowledge that in the following year Crome produced another painting entitled *View of Mousehold Heath near Norwich*.

The Norwich School painters of the early nineteenth century made much of the rural landscapes of East Anglia. Painters like John Crome, John Sell Cotman, Joseph Stannard and Henry Bright all celebrated the unkempt rusticity of the Norfolk and Suffolk countryside. Adhering to a set of commonly held goals, they resisted the urge to tidy up or glamorise the landscape in the way that those commissioned to portray it for the ruling aristocracy, who owned much of it, felt obliged to do.[2] Instead, these artists – largely working class and self-taught – celebrated the ordinary and commonplace: the labourers at work in the fields; the ponds, hedges and rivers; the rough, muddy, wheel-breaking tracks that threaded between the villages close to Norwich. Rather than a depiction of the 'improved' vision that landscape that architects like Capability Brown and Humphry Repton brought to the estates of the landowning class, this was a countryside that you could smell and taste. There was little that was Classical or Italianate about this raw, unimproved landscape, and little that might have been influenced by partaking of a grand tour. Rather, it was an honest portrayal of a working countryside that looked untidy yet beautiful and smelled of hay, human sweat and cow shit.

Crome's study is a rare representation of unadorned flint in visual art but there are a few other examples. The subject matter of Thomas Hennell's 1937–41 painting *Landscape: Flint heap, Road-making* is exactly as its title claims: a wintry landscape of a yet-to-be constructed country road; the flint, roughly broken and piled in a heap for use as hardcore. Hennell, who was primarily a watercolourist, was a contemporary of Eric Ravilious, who captured the pale chalky beauty of his native South Downs better than any other artist. Like Ravilious, Hennell enrolled as a war artist at the outbreak of World War II, and it was Hennell who was sent to Iceland to replace Ravilious after the artist died when his plane was lost off Iceland in 1942. Tragically, a similar fate also awaited Hennell, who would also die on active service four years later whilst in Indonesia.

An earlier war artist, Paul Nash, whose apocalyptic Flanders landscapes speak achingly of the futility of war, survived the ordeal but suffered from 'war strain' throughout the years that followed World War I. After the war he lived at Dymchurch on the Kent coast for many years and became fascinated with portraying the megaliths, stone circles and beech clumps of the nearby South Downs. During this period Nash's work became increasingly more expressive and abstract. His *Landscape of the Megaliths*, inspired by visits to the henge-encircled village of Avebury, resembles more a flint nodule than a landscape of sarsen stones. Later, in the 1930s, his work began to explore the potential of a life force, which he called a 'personage', within inanimate objects like the flints, bones and driftwood that he collected from Dorset beaches and assembled into small groups to paint and draw. His 1937 watercolour, *The Nest of Wild Stones,* along with a written piece of the same name, took inspiration from a black and white photograph of a group of flints made with his friend John Piper the previous year. It should be remembered that flint sometimes naturally takes the shape of a nest-like structure in the form of a paramoudra. It is the sort of nest that you might imagine a small dragon laying a clutch of eggs in.

*

Flint in its natural form, with its spiky intrusions and holes – the three-dimensional tracing of ancient burrows – is both abstract and organic. Its proclivity for forming fantastically shaped nodules has provided inspiration for modernist sculptors like Henry Moore and Barbara Hepworth. Moore, Hepworth and their friend Ben Nicholson were members of a group of artists who used to holiday together in the 1930s at a farm near Happisburgh on the Norfolk coast. A 2009 exhibition entitled *A Nest of Gentle Artists in the 1930s,* staged by both Norwich Castle Museum and Graves Art Gallery in Sheffield, took a closer look at these shared times on the Norfolk coast. The flyer for the exhibition shows a black and white photograph of six artists on Happisburgh beach with the cliffs forming a backdrop. The scene is a sartorial miscellany: Henry Moore and Ben Nicolson are both shirtless, while Barbara Hepworth, who eyes the camera defiantly, cigarette clenched firmly in her mouth, wears a black swimsuit and trousers. Standing to the left of the group, Ivon Hitchens and Irina Moore are dressed more formally, considerably overdressed for the occasion. Hitchens, who is carrying a raincoat and wearing a beret, gives the impression that he might have just arrived on the scene. The photograph was taken in 1931 by Douglas Jenkins, and it is Jenkins' wife Mary, seated next to Ben Nicholson, who makes the sixth member of the group. Photographs from the following year show some of the artists in far more informal dress, or rather, none at all, as they frolic naked on the beach. There must have been something about the Norfolk sea air that spurred on their convention-flouting Bohemian side. It also encouraged love and extra-marital dalliance, as Barbara Hepworth, whose husband John Skeaping had remained in London to begin with, would soon fall in love with Ben Nicholson, whose own wife had remained at home to look after their three children.[3]

Happisburgh was clearly an inspirational place for artists, sculptors especially, and Moore, Hepworth and Skeaping were all fascinated by the pebbles they found on the beach. Moore was equally interested in the pieces of weathered ironstone he came across. Skeaping recalled, 'Henry, Barbara and I used to pick up large iron-stone pebbles from the beach which were ideal for carving and polished up like bronze.'[4] Ideas also came from the flints Moore found here on the shoreline in Norfolk, and close to his studio in Hertfordshire. As a raw material, flint was impossibly hard to carve, but the intriguing biomorphic form of some of the pebbles proved to be inspirational: 'During visits to the Norfolk coast I began collecting flint pebbles. These showed Nature's treatment of stone and the principle of the opposition of bumps and hollows.'[5]

Henry Moore was already familiar with Norfolk and its coastline. In 1922 his parents and sister Mary had moved to the village of Wighton, a little inland from Wells-next-the-Sea on the north Norfolk coast. Moore was a student at the Royal College of Art at the time but spent his weekends and holidays at Wighton working in the garden of the family home. It was during this period that he first began collecting stones and shells from the nearby beaches.[6]

Wighton still has artistic connections to this day, and the village's All Saints Church is now the venue for the annual North Norfolk Exhibition Project, which was formerly staged at nearby Salthouse and Cley next the Sea. Like both those coastal villages, Wighton is a very flinty place. The village's church, cottages and

farm buildings all give the impression of having risen fully formed from the chalky subsoil, flint pebbles adhering to their walls on the way. A place faintly redolent of cow manure and silage, the village seems a little earthier than its coastal neighbours; not unkempt but less obviously gentrified and more grounded in traditional rural life. What was particularly noticeable when I visited the art exhibition at the church in late July was the vast number of swallows. The birds were everywhere, darting between farm buildings and gathering in rows on telephone wires. Even swifts, which had largely departed England for Africa, seemed to linger longer here.

Set a little way back from the high street is the old schoolhouse. This is now home to the Alfred Cohen Museum that celebrates the work of the American painter who had his studio here for many years. The artist's widow, Diana, is a trustee and can often be found showing guests around. A plaque embedded into the flint pebble by the door reads: *Alfred Cohen, Painter, lived and worked here 1978–2001.* Another plaque of similar design is inscribed with the legend: *Henry Moore OM, Sculptor, lived here 1922–1925.* Before its conversion to an art studio by Alfred Cohen, this was the home of Moore's parents and sister Mary, who was employed as head teacher at the village school next door. Half a century later, when the Cohens were clearing the overgrown garden in the early 1980s, the couple came upon an unfinished marble sculpture among the weeds – one of Moore's early attempts at a *Mother and Child*, a figure that would become a recurring theme for the artist.

Later in Moore's career, influenced by fellow artists like Nicholson and Hepworth, the sculptor's work would become softer, more rounded and increasingly abstract. Like Hepworth, his pieces would also feature holes and absent spaces. As Moore explains, 'The first hole made through a piece of stone is a revelation. The hole connects one side to another, making it immediately more three-dimensional. A hole can have as much shape-meaning as a solid mass.'[7] Moore's 1938 piece *Recumbent Figure* exemplifies this – a near abstract human figure incorporating a large hole that separates the head and shoulders from the rest of the body. The piece looks so organic and fluid in form that it might almost be a particularly intriguing hagstone.

Flint forms clearly became a strong influence on the sculptor, a fact confirmed by the archives held by the Henry Moore Studio at Perry Green in Hertfordshire. The way in which Moore transformed natural forms into three-dimensional works of art is evident from the preparatory drawings he made. Found objects like seashells and flints are scrutinised and drawn from different angles, their shapes metamorphosed to gradually assume human form. The catalogue of the Henry Moore Foundation lists numerous drawings made for later sculptures where the connection to flint is explicit: *Flint Torso* (1978), *Flint Torso Figure* (1980), *Flints and a Shell: Ideas for Sculpture* (1983), *Flint: Idea for Reclining Figure Sculpture* (1980), *Flint for Maquette for Spindle Piece*, *Flint for Maquette for Seated Figure: Arms Outstretched*, *Reclining Figure: Flint* (1977).[8]

In 2014 Moore's bronze three-metre-tall *Large Spindle Piece* was installed outside King's Cross station in London. This was the sculptor's own piece *(0/6)*, one of '6 + 1' bronzes of the work on display. On a five-year loan from the Henry Moore Foundation, the sculpture's selection was partly due to a consideration that its form suited the bold vertical lines of the station architecture. (Another deciding factor was that it was impossible to sit on or climb – certainly, choosing one of Moore's reclining figures instead would have invited trouble in this hectic and occasionally near-chaotic location.) The original design for the bronze was inspired by a knobbly flint picked up close to Moore's home studio. By a process of the abstraction and scaling up by means of maquettes and fibreglass models, the original form mutated and enlarged from a 20-centimetre piece of flint to a monumental three-metre-high bronze. An even larger version in marble adorns the lobby of a prestigious Miami hotel.[9]

Barbara Hepworth's work was influenced by flint in much the same way as Moore, especially with the inclusion of non-representational holes and voids in her sculpture. A 1932 sculpture, destroyed during World War II, called *Pierced Form* was an early expression of the inspiration garnered at Happisburgh. One of her later bronze works, the 1964 *Sea Form (Atlantic)*, has long been familiar to me as it used to stand next to the Whiffler Theatre, an outdoor performance area in Norwich Castle Gardens. Rather than the flints of East Anglia, the piece was influenced by the standing stones of Cornwall, where Hepworth had her studio. The sculpture has the overall form of a Cornish menhir, but it is pierced by holes and openings at various heights. While the group of Cornish megaliths known as Mên-an-Tol has a stone with a circular hole at its centre, the hard granite of Cornwall does not naturally harbour holes in the way that flint nodules do, so it is likely that there was still the memory of Happisburgh beach hagstones present in its design.

In recent years Barbara Hepworth's *Sea Form (Atlantic)* has taken up residence outside another Norwich theatre, the Norwich Playhouse, close to a bridge over the River Wensum. I have a black and white photograph, taken at the original 1970s location, that has my friend Martin peering wistfully through one of the sculpture's head-height holes. Reminded of this, I got him to pose again when he visited Norwich in 2022. Apart from new location – the sculpture had been moved a few hundred metres in the interim – it appeared unchanged, unperturbed by nearly half a century's exposure to the elements. Martin looked much the same too, albeit a little greyer than in the earlier photograph. Stones age slowly, very slowly, while bronze attains a wholesome greenish patina over time. Unlike human ageing, it is all surface.

The sinuous form of a flint nodule is not always used so abstractly in art. Sometimes it can become the subject of the work. I am thinking of the sculpture *Man of Stones* by the Suffolk artist Laurence Edwards, a work I came across unexpectedly one day whilst walking on the campus of my alma mater, the University of East Anglia. UEA Broad, the body of water that lies south of the university's concrete teaching blocks and distinctive ziggurat-shaped halls of residence, has a path that runs along its southern shore close to the River Wensum. One day, while retreading the all-too-familiar route after an absence of many months, I found myself confronted by the newly installed statue.

The figure, a naked two and a half metre-tall male, is set back slightly from the path half framed by reeds. Constructed of bronze, it has been roughly rendered to give the impression of a figure rising from the ground caked in mud. Suspended around the figure's neck is an abundance of stones on ropes. More accurately, these are large flint pebbles and hag stones that have been threaded onto the ropes to make burdensome necklaces that resemble shackles.

Seeing it for the first time, there was something that struck me as disturbing and slightly fetishistic about the sculpture: a strong male figure that appeared to carry his rocky burden with considerable dignity. It seemed to embody suffering and stoic forbearance. Reading the small plaque, I was also taken aback by the artist's name: my first and middle names with the single addition of an 's'. It all felt disarmingly personal – the flint, the name. Was this just another example of the Baader-Meinhof phenomenon working to grab attention?

With downcast eyes, *Man of Stones* faced out across the broad to the long concrete terrace of the university buildings beyond. Silhouetted in the distance on the terrace roof were two more human forms: unmistakable Gormley figures. I remembered the controversy about their original installation here. The cast iron figures, part of Sir Antony Gormley's nationwide *Another Time* project, had been at the university since 2017, installed 'as a striking and thought-provoking addition to the campus for our students, staff and visitors'. From the outset there had been calls to get rid of them as some considered the sculptures to be in poor taste. One objector ('Anonymous N') even started an online petition to remove the figures, but this received little support.[10] Gormley himself refuted all accusations and insisted that they were 'nothing to do with suicide', asserting that all art should be unsettling in some way.[11]

Man of Stones was unsettling in a different kind of way. The sculptor had conceived it as a figure emerging from the mud, something specific to the watery location it was placed in. Originally modelled in clay before being cast in bronze, Edwards collected source material – twigs, branches and reedmace – from the UEA Broad site and these were incorporated into the figure. The flints that the figure carried around his neck, although on first glance giving the impression of a burden, were intended to symbolise the ground from which he had emerged: a connection between humanity, geology and landscape. The intent, rather than provoking discomfort, is that the figure is seen as a guardian who offers protection to its watery home: 'less a haunting presence and more of a blessing.'[12]

In a film by Bill Jackson that follows the conception, building and casting of the figure, Edwards talks about his thoughts behind the work: 'My process has always been about loss; my process is really about losing things…about dissipation.' He describes the lost wax process in which the clay figure disappears to become solid again, reincarnated as a wax man before this, in turn, is transmuted into metal. '*Man of Stones* comes from the idea of accretion and idea of being out of your own environment, the idea of gathering objects and making your environment from your immediate place.' He reflects too on the importance of landscape, his gut response to it, and the fragility and impermanence of humanity and the geological history of the landscape we live in: 'working with flint, for example, is very important.' The stones for the sculpture were gathered from beaches on the Suffolk and Norfolk coastline, Covehithe on the Suffolk coast being one specific location. Edwards selected those he thought most resembled human organs like lungs, kidneys and bladders: 'They're visceral, intestinal. When you cast them they begin to lose their stone-ness and begin to look and feel like human organs.'[13]

The lost-wax process echoes the way that flint is formed in nature. The human invention of allowing molten metal to occupy the void left by melting wax parallels the way that liquid silica fills the space abandoned by burrowing marine creatures in chalk sediment. The connection goes deeper. The lost-wax technique of casting metal is the second oldest technology known to mankind, a knowledge that goes back thousands of years to the early Bronze Age. The oldest technology – more ancient still – belongs to flint itself.

*

The presence of flint denotes the absence of something else: flint is a relinquished space filled by precipitated silicon from sea sponges and siliceous plankton. In its gelatinous silica form, flint occupies voids in the sediment: the cavities left by burrowing crustaceans and molluscs. Often it envelops Cretaceous species like sea urchins to leave trace fossils. Flint's curious nodular form may also result in accidental simulacra, with shapes that resemble animal parts like ape jaws and dinosaur claws. Some may even possess the sort of weird anthropomorphic forms that so entranced artists like Moore and Hepworth. Others, like those chosen by Laurence Edwards for *Man of Stones*, may even bear a resemblance to our internal organs. Flint can be visceral, quite literally.

An artist who is especially interested in the space defined by absence is Rachel Whiteread. Whiteread is probably best known for her 1993 piece *House*, which until its demolition stood on the site of former Victorian terrace house in East London. *House* was created by filling the whole interior of the original house with concrete before removing its brick fabric to leave a cast of the building's interior standing. In Whiteread's words, the work represented 'the residue of years and years of use'.[14] Space made solid, an absence given corporality: to say *House* was controversial is to underplay its effect, as the work won both Turner Prize for the best British young artist in 1993 and also the 1994 K Foundation award for the 'worst artist of the year'. For others, in particular city planners and developers, criticism was more of a political nature. Whiteread's work simply did not gel with the manufactured zeitgeist of the increasingly gentrified, upwardly mobile East London of the 1990s. Always envisaged as a temporary structure, House was destroyed and removed the year after its creation.

Given the nature of the concrete mould, a working-class dwelling replete with the thoughts, feelings and memories of those of who had lived there, *House* was as much memorial as it was artwork. It was a monument to the interior life the house had experienced through the years, a concrete repository of bygone East End life. It was also a fragment of the historic built environment of the locality, an area of soon-to-be redeveloped East London in sight of Canary Wharf.

House recognised that a domestic dwelling is more than real estate, more than an architectural item. Although not intentional, Whiteread's use of a manmade structure as a casting mould was an echo of the way that flint in the form of liquid silica came into being by occupying voided space. A house is a home, a space we fill with the comforts of domesticity. For the French philosopher Gaston Bachelard, a house represents in phenomenological terms a space where 'our consciousness of well-being should call for comparison with animals in their shelters.'[15] A house/home is a hearth, a haven, a place of safety and nurturing; a nest. An artists' colony might also describe itself in the same way, as did the somewhat self-consciously titled 'nest of gentle artists' in pre-war Happisburgh.

It is interesting to note that Whiteread's favoured medium is concrete. Concrete is ubiquitous in the modern world. The ultimate go-to material for fast, cheap and reliably strong construction, concrete is DIY geology. There can be nothing that is manmade in this world that has greater overall mass. Addicted to building, humankind seems intent to hard landscape as much of the Earth's surface as possible, banishing nature at every turn. The irony is that concrete is created from naturally occurring rock in the first place: a concoction of sand and gravel combined with cement derived from limestone.

*

There is one more artist to mention. In 2022, to coincide with the Norfolk & Norwich Festival that year, Norwich Castle Museum announced an exhibition titled *Ode to a Flint*. Given my interest in flint its timing felt quite opportune, serendipitous even. The exhibition was by Yelena Popova, a Nottingham-based artist of Russian origin.

When I went along to see it the museum was undergoing extensive renovations and consequently receiving very few visitors. I arrived to find that the room where the exhibition was to be held had a sign proclaiming: 'Withdrawn in Solidarity #No War'. In lieu of the exhibition there was a small installation of flints from the museum's own collection along with booklets describing what had originally been planned. There had been a request to bring along a flint to contribute to the installation and so I did, placing a small Norfolk flint among the larger, more impressive ones that had come from the museum's extensive collection. The booklet explained that in December 2018 Popova had visited a beach near Sizewell nuclear power station on the Suffolk coast. Here she had found a flint that marked the beginning of a new body of work that looked at 'connections between the UK nuclear industry, landscape and the passage of time.' The resulting exhibition *Ode to a Flint* had been planned as an invitation to consider complex geological histories through painting, tapestries and installation.[16]

In the wake of the cancelled exhibition, I contacted Yelena Popova to tell her about my own interest in flint and to learn more about her work. She responded to tell me that the *Ode to a Flint* show was intended as a version of the *Scholar Stones Project* she had produced for the Holden Gallery in Manchester in 2018. The goal of the Norwich exhibition was to bring what might be considered local 'scholar stones' – Norfolk flints – to view. Norwich Castle Museum has a large geological collection but generally its flints are not exhibited. The planned exhibition would have included some of the museum's collection alongside the artist's own soil paintings and tapestries.

Scholar stones were originally a Chinese art tradition from the Tang dynasty, a school of thought that later found expression in the Japanese *suiseki* and Korean *suseok* traditions. A scholar stone, or *gongshi*, alternatively known as a 'spirit stone', is a naturally occurring rock that, chosen in accordance with a set of four important qualities, becomes a valued artefact that can be placed in a garden or stand freely on its own. Traditionally of limestone, they are considered as a microcosm of a larger landscape:

an aesthetic in which nature makes art in its own image as a fractal of itself. The human element is as much curatorial as anything else and involves the appropriate selection of rocks that are observed to contain a miniaturised landscape within. While chunks of weathered limestone easily lend themselves as stone-sized representations of the wider landscape, flint is different. Flint reflects an altogether different world that demands a different aesthetic. An individual flint can never be a simulacrum of the landscape it is found in. Each flint is its own world, a micro-landscape of holes, conchoidal scoops and bulges. When it is fractured it behaves fractally – each resultant shard is a new, almost grain-free, sharp-edge fragment that can be further reduced to a miniature version of itself.

During the Covid lockdowns Popova had been running a 'scholar swap' scheme on Instagram in which she exchanged stones foraged in her travels for books. In one email exchange we agreed to swap one of her stones for a copy of one of my books. The stone arrived a few days later, solidly enclosed in brown cardboard and gaffer tape. In true *gongshi* tradition it came complete with a small wooden display base. It was indeed a world in miniature, a microcosm of secrets. Partly encased in a cortex that was stained with the grubby patina of time, the stone was randomly punctuated with holes and had pinprick indentations that suggested the ancient sponges to which it owed its existence. I was told that it probably came from Sizewell on the Suffolk coast, although this was not certain. It did not matter. It felt good to be gifted a flint that I had not chosen myself. Here was my own scholar stone, my silica spirit object.

The Scholar Stones Project, like most of Popova's work, is multimedia and involves the use of found objects – 'scholar stone' flints and paintings made using soil gathered from decommissioned nuclear sites – and woven tapestries. Each element carries what Popova calls the 'geological imprint' of the site. Tapestries, like '*Keepsafe*', are offered as plans for mausoleums to house the decommissioned Magnox reactors entombed beneath the nuclear architecture: woven propositions for tombs whose triangular form echo the Pyramids or perhaps even Lenin's mausoleum in Moscow's Red Square.[17]

What resonated with me most was the consideration of the contaminated graphite within the reactor and the carbon-14 it harboured among a miscellany of radioactive waste. In scientific terms, carbon-14, with its unique radiometric ability to date decaying organic matter, is the key to the distant past. The half-life of carbon-14 is approximately 5,730 years: roughly the same length of time that, in northwest Europe at least, connects Neolithic man with his post-nuclear, Anthropocene descendants: a span of human life that ranges from the technology of stone to that of nuclear power.

Radiocarbon dating was developed in the late 1940s by Willard Frank Libby, a physical chemistry professor from Chicago University. Based upon the earlier work of Martin Kamen and Sam Reuben, who discovered the C-14 isotope in 1940, the scientific innovation went on to win Libby the Nobel Prize for Chemistry in 1960. The adoption of carbon dating marked a huge step forward for field archaeology as it enabled the accurate dating of organic artefacts that were between around 7,000 and 50,000 years in age. A notable irony is that both the development of carbon dating and the capability for nuclear fission came from the researches of the Manhattan Project, with which Professor Libby was involved during World War II. The project spawned two clear outcomes, interconnected but diametrically opposed in terms of the human condition. One provided a way of better understanding the past; the other was a means, if we are not careful, of obliterating the future.

THIRTEEN
CONTEMPLATING DUST

Norwich – Norfolk | Coventry – West Midlands
Kirby Bedon, Surlingham, Shotesham, Thetford, Reepham,
Limpenhoe, Wickhampton – Norfolk

Where there is ruin, there is hope for a treasure.
Rumi

The solitary church tower that gifts its name to St Benedict's Street in Norwich almost gives the impression that it had always been meant to stand alone, as if there had never been a body attached to the octagon-topped round tower that was the head. Finding fresh purpose as the centrepiece of a small housing scheme in the 1980s gives the tower a dignity that surpasses any notion of a folly, or worse, a ruin. The pile of rubble that was once the nave, destroyed overnight by Luftwaffe bombers in 1942, has long been swept from memory and removed from the map by urban planners.

No one likes reminders of the savagery of war, even if there are some who argue that the destruction that results from it should remain visible as some form of cautionary sign in times of peace. It is quite probable that the flint rubble of St Benedict's bomb-blasted eleventh-century walls has since been recycled as hardcore in some post-war development scheme. It may even lie beneath the 1970s inner ring road, which sweeps within earshot just beyond the broken-toothed remains of the medieval city wall – another ruin, albeit one that is more the product of neglect and urban scavenging than armed conflict. Civic pride and understandable concerns about health and safety usually result in the ruins of the urban environment being tidied up in this way, and if not tidied, then boarded up to keep the public out. But you cannot realistically board up a pile of rubble, and to put a fence around it is a solution that is both impermanent and relatively high maintenance. For those inclined, a protective fence also serves as an invitation to view. Nothing quite says 'Come and have a look' better than a notice that snarls *KEEP OUT*.

In the countryside things are different. Semi-ruined and abandoned buildings – churches especially – can easily outlast their apparent sell-by date. Given sufficient time, ruins can blend into the landscape and accumulate folklore along with the ivy and bramble. 'All ruins are precious' asserts John Burnside in his essay 'Aurochs', before going on to point out that they should be considered as both healthy and regenerative: places that thrive with the plants that colonise them, the bees that feed on the blossoming ivy and honeysuckle, the nesting birds. Ruins signify continuity, a visible connection to the past. They make us consider the brevity of the human lifespan and the hubris that seems to be an inevitable by-product of civilisation. Their material incompleteness becomes part of a place's history; a degree of ruination adds character and mystery.

A ruin invariably provokes a sense of melancholy – a psychological linkage of place and emotion that has been recognised since antiquity. There is even an Old English word for it: ***dustsceawung***, which translates as 'the contemplation of dust', although 'dust' here should be considered in the broader sense of that which remains after destruction, along with the concomitant awareness that all things go this way eventually. Ruins may also evoke fear and mystery. Images of silhouetted skeletal remains frequently find their way onto the book covers of Gothic novels and folk horror potboilers.

Given the correct political spin, ruins can also become a means of expressing resilience in the face of brutal attack. Picasso's *Guernica*, a vivid portrayal of the horrific 1937 bombing of a small Basque town by Nazi Germany and Fascist Italy at General Franco's request, remains a powerful symbol of vicious oppression and serves as a rallying call for the fight against fascism. It was in this spirit that, during World War II, the War Artists Advisory Committee (WAAC) recruited artists like John Piper and Graham Sutherland to paint the devastation wreaked by the Luftwaffe in British cities. The pictures were intended as both a record and a boost to morale.[1] Piper's first employment was in Coventry, which was near flattened by the payload of 400 German bombers on the night of 15 November 1940. The following morning Piper painted the fractured stone tracery of the city's St Michael's Cathedral apse while the building was still burning: a visceral representation of the pain felt by both the artist and the ill-fated victims of the Coventry raid. Like all Piper's paintings of architecture, the work was powerful, employing hyper-real colour chosen from a spectrum that represented and intensified the building's inner spirit.[2] It was also undeniably, and disturbingly, beautiful. In later years, both Piper and Sutherland would be involved in embellishing the bold new cathedral that would rise phoenix-like on the same site. The remains of St Michael's, those burning skeletal ruins that Piper painted on the morning after the raid, would remain in place as testament to the horrors of war.

After the war, Piper wrote a somewhat controversial essay entitled 'Pleasing Decay', in which he argued that ruins, if structurally sound, should be kept as they are and conserved for their picturesque qualities. The writer and traveller Rose Macaulay, writing *Pleasure of Ruins* a few years later, makes a similar point.[3] While both argued for an aesthetic that embraced ruin – classical or medieval – for the intrinsic beauty contained within the decay process, it should not be forgotten that both were written at a time when the modern ruins of war were still very visible on the ground. Neither was oblivious to this. Both had experienced war's destruction first hand: Piper at Coventry, Bath and other bombed cities; Macaulay in London, where her flat and its entire contents had been destroyed in an air raid. To admire the slow-shifting beauty of an abandoned building left to colonisation by nature is one thing; to relish violently inflicted war damage as a sort of visual *Schadenfreude* is another.

*

Norfolk, a county whose countryside occasionally hints at an undercurrent of feral neglect, has more than its fair share of ruins. In particular, it abounds with a wealth of long-abandoned flint-built churches. Mostly these ended up as ruins because of abandonment and their subsequent deterioration over the centuries that followed. Others were deliberately dismantled, partially at least for the building stone they held, which would then be recycled for use in new churches, houses and farm buildings. External factors would also sometimes come into play. While wartime bombing raids were responsible for the destruction of several Norwich churches, most notably the shrine church of St Julian's and the aforementioned St Benedict's, damage could sometimes be wrought from the air even in peacetime. In 1947 the tower of the already decaying twelfth-century church of Tivetshall St Mary in south Norfolk was brought down by the vibrations caused by the sonic boom of an overflying jet aircraft.

The church, now evocatively overgrown with ivy and bramble, remains standing as one of many isolated ruins in the county. Each one is a ready-made ghost story awaiting an audience.

Not far from Norwich in the village of Kirby Bedon is the ruin of St Mary's. Standing next door to the still-active church of St Andrew's, St Mary's has been in a ruinous state for centuries. A drawing of 1814 depicts its nave and tower still erect, although covered in ivy and clearly abandoned. The ruin underwent more rapid depletion in recent years when, in February 2020, an estimated four tonnes of flint fell without warning from the tower. It must have been a relatively muted collapse, as the congregation attending Sunday Service in neighbouring St Andrew's at the time of the fall was completely unaware of what had happened.

Visiting the village later that same year I found the ruin and rubble fenced off for reasons of public safety. Even so, it was easy enough to sneak through to have a look. Prior to the latest fall, St Mary's had been a pleasing ruin, a gloomy ivy-clad, broken-toothed tower that hinted at Gothic horror – the classic jackdaw hi-rise. Now its height had been truncated significantly and what remained of its belfry had tumbled to the ground to leave a single precarious fang of crumbling flintwork. The tower's circular interior was filled with a hefty pile of flint rubble and lime mortar – medieval architecture deconstructed to its base elements. The large flints that lay heaped on the ground had probably originally been cleared from local fields and put to good use in building the tower: no waste, no energy-sapping stone-miles involved. Like many flint churches in the region, the construction process amounted to little more than a vertical rearrangement of a horizontally inclined landscape. Now the flints had fallen back to Earth, finally succumbing to gravity after seven long centuries aloft.

The village of Surlingham, down the road in the Yare Valley, has another St Mary's: a round tower church that is still active and in altogether much better shape. Just a few hundred metres away is the twelfth-century ruin of St Saviour's, a church that became redundant when the village's two parishes were consolidated in the eighteenth century. St Saviour's is an evocative ruin that has two gravity-defying arches suspended above its broken walls. With the exception of a few well-placed bricks in the arch, the church is built almost entirely of flint, and even the quoins at the corners are of roughly squared flint blocks. The attractively overgrown site sits on a low rise that gives a good view over the River Yare as it flows east towards the coast at Great Yarmouth. It is a peaceful spot of buzzing bees, mooing cows and the distant chug of river traffic. Outside the low fence that encloses the ruin are the modest grave markers of a local couple: the celebrated naturalist Ted Ellis and his wife Phyllis, who lived nearby at Wheatfen Broad. Along with a simple plaque, both of the graves are marked with paramoudra, which were presumably obtained from the same source as the black flint used in the church.

When I posted a photograph of St Saviour's on Twitter, more as an aesthetic statement than any sort of query, a local geologist, Tim Holt-Wilson, responded with: 'It has monster flint boulders used as quoins – quite unusual. I put it down to the proximity of the Norwich Crag basement bed in the Yare valley side, with its bullhead nodules.' I wondered how unusual this was, and how widespread 'bullhead nodules' were. Were the flint boulders local? 'Yes', Tim responded, 'The Crag basement bed resting on the Chalk, has outcrops in quarries along the Yare valley side, as at Whitlingham, Kirby Bedon and Bramerton.'

Ruined churches can be found all over Norfolk. The village of Shotesham in the south of the county has three standing churches, two of which are active and the other ruined. All Saint's stands on a low hill with a fine view of gently undulating countryside, while the village's other church, St Mary's, is at some distance away atop another low rise. Down the lane from St Mary's is the ruin of St Martin's, now little more than a tower alongside the low walls of a roofless chancel and nave. A matrix of ivy enveloped the ruined tower until recently but this has since been removed to prevent any further decay. A photograph, selectively framed, can easily bring the two similarly sized square towers into juxtaposition as a sort of 'before and after' capture of architectural decline. Both churches are, of course, built largely of knapped flint.

The neighbouring village of Saxlingham Nethergate has another St Mary's, as does the bordering parish of Saxlingham Thorpe. While the former, with its clock and sundial-bedecked tower, stands bright and proud in a neatly mown graveyard surrounded by mistletoe-strewn poplars, a centrepiece to what looks like a peaceful, well-heeled village, the latter is tucked away in the deep shade of a pocket of woodland a little way outside Saxlingham Nethergate. The village this St Mary's formerly served has since long vanished from the map, and the ruin would probably remain largely hidden from public gaze were it not for the footpath that runs alongside it. All that remains are low flint walls, carpeted with moss, and a shattered square tower that has little more than a single side remaining. As elsewhere, much of the church's original flint has been carted away to see service as building repair and hardcore.

The historical record is incomplete but it appears that the church was abandoned by 1687, although the building might have seen some sort of agricultural use in the centuries since. Certainly, some of the flint was taken to repair its sister church in Saxlingham Nethergate. Deserted, the settlement around the church no longer exists, not even as a historic site symbol on the OS map. The original settlement, like another two hundred or so in Norfolk alone, vanished from sight as villagers upped sticks to more hospitable locations. Some of the villagers probably migrated to the west of the parish by the River Tas, where there was a mill and the main Norwich to Ipswich road. Others moved a little way east to the hamlet that is now known as Saxlingham Green. The question is: what was so inhospitable about the original location? Despite the common and often erroneous assumption that medieval abandonments are inevitably the result of plague, it was more likely a matter of exhausted soil or problems with drainage on heavy clay soil. For centuries, this ruin in the wood had been a peopled place, a focal point for the community. On medieval Sundays, St Mary's Saxlingham Thorpe would have thronged with people, with farm workers coming together for the only respite in their long working week. It would have been an occasion to dress up in best, to sing and pray, to tuck forelocks at the gentry and eye up the local talent. These ivied walls, once plastered and covered with biblical imagery, would have echoed with the singing of hymns and telling of sermons. They would have witnessed both the sobbing of funerals and the more joyful outpourings that attended happier occasions like weddings and christenings. Now, half-collapsed, buckled under the weight of neglect and time, it is only the moss-covered flints of its truncated walls that retain any memory of what once took place here.

*

While some ruined flint churches are tucked away from sight, others stand in plain view amidst the brown corduroy of the ploughed fields, vestiges of medieval stone, stark and solitary against the wide sky. Many, like St Mary's at Saxlingham Thorpe, are modest in size, the medieval remains of an abandoned parish church. At other places, like the Norman-built priories at Thetford in the Norfolk Brecks or Castle Acre in the west of the county, abandoned flint can be witnessed on a much grander scale.

In recent times Thetford is best known for its extensive council estates that in the 1960s housed what used to be termed 'London overspill', its largish Portuguese population, and its historical association with the Sikh Maharajah, Duleep Singh. It was also the birthplace of the revolutionary Thomas Paine, a man who had a part to play in both French and American revolutions, yet, in his own county at least, never achieved the fame or the reverence received by that more gung-ho Norfolk-born hero Horatio Nelson.

While today it may appear slightly downbeat and firmly locked into the late twentieth century, Thetford is an ancient town with a rich cultural heritage. A town that according to the poet Edward Thomas was 'a place sometimes burnt, sometimes fortified by the Danes',[4] Thetford had been an Iceni stronghold and important Saxon town long before the Norman Conquest and had possessed both a monastery and a grammar school prior to 1066. In adherence with standard Norman practice, a castle was built immediately after the Conquest, and shortly after this, in 1103, the Cluniac Priory of St Mary was founded in the town by the knight Roger Bigod. The priory would go on to become one of the most important religious institutions in East Anglia and, thanks to an appearance of the Virgin Mary in the thirteenth century, it later became a pilgrimage site and remained so until the Dissolution of the Monasteries nearly two hundred years later. Today the whole priory complex is a ruin, although some elements like its Grade I listed fourteenth-century gatehouse remain almost complete.

I decided to visit Thetford to take a closer look at its flint connections. My first port of call was the town's Ancient House Museum, where upstairs I found a small display on the Neolithic flint industry at nearby Grime's Graves. A cabinet that focused on axe identification bore the caption: *Is it a flint tool?* There was also a large piece of un-worked floorstone, the highest quality flint found in the mines. Beneath the thin cortex that coated it like a plaster of Paris skin, the stone was the deepest of blacks – darker, even, than the knapped flint I had seen in the walls of some of the town's buildings. Downstairs in the video display room, easily overlooked in its modest wooden frame, was an alphabet that had each letter A to Z shaped in flint. Handmade in Brandon in the 1920s by Bill Basham – a fitting name for a flintknapper – the accompanying description said that it took two years to complete. This was not surprising, as the work must have required enormous skill and patience. Flint, which fractures in a way that does not follow natural planes of separation, does not easily lend itself to be shaped into abstract combinations of curves and straight lines. Bill Basham probably made this simply because he *could*, because he was in possession of all the requisite skills. More than a voluntary apprentice piece – the work, in fact, of a real master – this modest display of the flintknapper's art demonstrated a skill that went far beyond the normal call of duty.

A little way down the hill from the Ancient House is the Church of St Peter, its solid square tower looming high over the narrow street. A plaque at its doorway reads: *One of three surviving Thetford churches. St Peter's was extensively rebuilt in 1789–91 by public subscription. Sometimes known as the 'black church' due to the extensive use of black flint, which creates a chequer board pattern on the tower.*

Black it certainly was, although the effect of all the knapped flint seemed to create an impression that was dark and gloomy rather than anything bright and full of hope. I noted the date of its rebuild: the first years of the French Revolution, a time when the town's own resolutely anti-Church, anti-monarchy son, Thomas Paine, was living in France and deeply involved in revolutionary matters. Perhaps the 'public subscription' rebuilding of the church tower at this time was seen as an act of propitiation against the existential fear of similar things happening to England's own *Ancien Régime*. Meanwhile, at the same time over in France, Paine was busy writing *The Rights of Man* in defence of the revolution he was witnessing. Paine's gilded statue, sculpted by Sir Charles Wheeler, President of the Royal Academy, and erected on King Street in 1964, can be seen from the churchyard – black flint and gilded bronze, Protestant establishment and nonconformist republican Deist, brought together in the sweep of an eye.[5]

Thetford Priory ruins cover an area of grassy parkland at the edge of town, a popular place for local dog walkers to exercise their pets. I had been here a couple of times before but had forgotten quite how extensive the site is. Ruined or not, what remains is quite substantial, with the skeletal remnants of tall arches probing high above the priory walls. Even as a ruin, the sheer size of it is impressive – the extent of the grounds, the obvious grandeur of the original buildings of church, cloisters and monks' lodging. The surviving walls of the priory are largely devoid of their original Caen limestone cladding and the thick flint foundations that were originally intended to lie unseen are now fully exposed. The flint is of the same inky tone as that of the town's 'black church' of St Peter.

Thetford Priory had once been an institution of great importance. Following the Norman Conquest and the deposition of Egelmar, Saxon bishop of East Anglia, it was Thetford that was chosen as the new base for the bishopric (this was later moved to Norwich by Herbert de Losinga). The priory of Our Lady of Thetford was founded by Roger Bigod in 1107 in the reign of Henry I. As with the priory at Castle Acre in west Norfolk, founded around the same time, Thetford Priory was manned by Cluniac monks from Lewes in West Suffolk. As the burial place of the Earls and Dukes and Norfolk it thrived as an institution until the reign of Henry VIII, although even then it was one of the last religious institutions of its kind to be dissolved.

Thetford went into modest decline around the time as the priory's dissolution. The town had little to fall back on, as it was never involved in the booming cloth weaving industry that made other towns of Norfolk and Suffolk prosperous.[6] The surrounding countryside suffered similarly. Cleared of peasants, then enclosed for sheep and heavily overexploited, much of the Brecks ended up as an agricultural desert: quite literally, as the Brecks became the only region in the British Isles plagued with drifting dunes of sand thanks to profligate overgrazing.

*

Flints found in church walls are not always quite what they may seem. While building material was routinely recycled as churches were abandoned and allowed to go to ruin, flint sometimes travelled much further than simply between neighbouring churches. Those closer to the coast were especially liable to have a surprisingly varied geology embedded in their walls.

I signed up for a day course that examined the stonework of a group of churches in the Norfolk Broads: St John the Baptist at Reepham, St Botolph's at Limpenhoe and St Andrew's at Wickhampton. While each of these churches was constructed largely of local flint, their walls also contained fragments of rocks that were clearly alien to the region. Mostly these were pieces of sandstone and dolerite that had been borne by glaciers from upland Britain and used opportunistically when the church was built. The Reepham church was particularly diverse in its use of alien rock and had walls that contained oolitic limestone and ferricrete conglomerate as well as locally sourced flint. These same walls also made use of any Roman material that was conveniently to hand in the locality. In Roman times, when Reepham had stood at the mouth of the wide estuary that is now Breydon Water, there had been a lighthouse, and the bricks that had been recycled for use in the walls of St John the Baptist almost certainly originated from what remained of this. Brick was not the only Roman material used. The walls of St John the Baptist also contained pieces of dressed carrstone, a ginger-toned sandstone native to northwest Norfolk that had probably been transported from the ruins of the Saxon Shore fort at Branodunum.

We looked at the coursing of walls and the use of brick and knapped flint, and discussed the techniques that might be used to date them. I learned that while locally produced bricks were used in construction from the early fifteenth century onwards, poor quality, uneven bricks from the Low Countries had been imported long before this, from the late thirteenth century onwards. These imported bricks often arrived here as ship's ballast. So did unfamiliar rock. East Anglia had strong trade links with northwest Europe throughout the medieval period, and one of the biggest exports was Norfolk and Suffolk wool. Scandinavian ships crossing the Baltic to load up with wool at an East Anglia port would often bring Baltic rocks as ballast in their hold, which they would offload before taking on their cargo. Some of these rocks – pieces of basalt, granite and breccia – would eventually find their way into the walls of churches and other buildings.

We moved on to St Botolph's, Limpenhoe, where Tim Holt-Wilson, who along with Ian Hinton was leading the tour, pointed out the way that the flint courses would often have a clear 'end of year' boundary that contained a thicker layer of mortar, an indication where work stopped in autumn before recommencing in the late spring of the following year. This was the result of only being able to build with confidence for about five or six months a year: any threat of frost would render the stonework unstable. Tim also drew our attention to the different types of flint used in the construction process. Other than those frost-damaged or plough-broken nodules that lay on the fields, or the rounded pebbles that could be gathered on nearby shores, flint lies deeply buried in the Broads region. Material of a quality good enough for knapping had to be brought in from further afield. The flint on show, stained by minerals and patinated by exposure to the elements, was surprisingly variable in colour and texture. For the inexperienced like me it was difficult to identify the stone's origin, but here Tim's considerable knowledge of local geology came into play. In the broadest terms, the flint that originated in the Norwich area tended to be a darkish brown colour, while that from the Brandon region of the Brecks was the deepest black – this, at least, was something that I had some knowledge of.

At St Andrew's, Wickhampton, Tim pointed out what he called 'chattermarked' flints: stones that came from storm beaches and

which bore tell-tale semicircular marks that were the result of attrition. Later, as I was looking at a large piece of flint set in the wall next to a weathered gargoyle figure, Tim came up and said, 'I see you've found a piece of Swedish flint.' To my mind I hadn't found anything of the sort, although the stone did strike me as being a little out of the ordinary. He pointed out the slightly grainy, matt texture of the surface – a clue to an origin that lay far beyond the parish, even beyond the horizon. This was a stone that had probably arrived as ship's ballast from the Swedish coast having already been transported by ice from another part of Sweden. A stone steeped in deep history, an orphan that had rested motionless in the chalk for aeons only to be shifted relentlessly – ice to shore, shore to ship, ship to harbour, harbour to church – all in the tiniest fraction of a hundred-million-year lifetime.

FOURTEEN
FLINT TIME

Clerkenwell – London | Hoxne – Suffolk
Isle of Arran, Siccar Point – Berwickshire

And I hold the flint, full of stars. And wonder.
Alan Garner

In 1679 the apothecary and pioneer archaeologist John Conyers attended the excavation of the remains of what was thought to be an elephant on the banks of the River Fleet in Clerkenwell, London. The location, known as Black Mary's Hole, was a healing well, which despite the supposed therapeutic properties of its water was considered a place of ill standing, having a reputation for debauchery and highway robbery; in later years it would gain notoriety as one of London's first recorded gay cruising grounds. Even today, the area around King's Cross carries a whiff of its supposedly insalubrious past, although now it is mostly concerned with its role as a transit hub and providing the quick-fix requirements of travellers. It is easy enough to cast the mind back a few hundred years and imagine the crime and corruption that once took place here. It is trickier, perhaps, to imagine elephants roaming the wastes of Clerkenwell.

The well and the River Fleet are now both long lost to view. The Fleet, which degraded into an open sewer during the late medieval period, has since been comprehensively culverted and its presence can now only be detected by placing an ear close to drainage grids at certain localities along the river's ghosted course. Names, too, have been redacted from the modern London WC1 map, and Black Mary's Hole is no longer marked. The former site of the well lies forgotten beneath new development somewhere in the region of Margery Street, east of King's Cross Road.

The elephant was initially thought to be an antediluvian pachyderm, a casualty of Noah's biblical Flood, although the antiquarian John Bagford would later suggest that the elephant may have belonged to the invading army of the Roman emperor Claudius. With the benefit of hindsight we can now be confident that neither appraisal was correct: the elephant was in fact far older. A tooth found intact at the excavation has since been shown to belong to that of a mammoth or straight-tusked elephant (*E. antiquus),* an extinct species that roamed Europe from the Middle to Late Pleistocene in the period between approximately 780,000 and 30,000 years ago.

A manmade object was also found close to where the elephant tooth had been discovered: a sharply pointed flint hand axe. Quite reasonably, the axe was initially considered to be of the same antiquity as the elephant, perhaps something that had been used by the beleaguered Britons in defence against the animal. It was necessary to invent a scenario that considered both tooth and axe to be coeval, as seventeenth-century religious convention required that prehistory fitted within a tightly constrained time frame, and the world at that time was thought to be only six thousand years old. More precisely, according to the calculations of the then recently deceased Bishop of Armagh, Reverend Bishop Ussher, the world and all its wonders was believed to have come miraculously into being at around 9am on 22 October 4004 BCE, a Sunday. Anyone who was wealthy or healthy enough to attain the biblically approved age of three score years and ten in the seventeenth century could have looked back on a life that, by Ussher's reckoning, would have spanned more than one percent of all worldly existence.

*

The Restoration period of the late seventeenth century was a time that saw the gestation and awkward birth of what would later become known as the Enlightenment. Science was in its infancy, or to extend the analogy, still stirring in the womb waiting to be born. Although the Renaissance polymath Nicolaus Copernicus had published his controversially heliocentric *De revolutionibus orbium coelestium ('On the Revolutions of the Heavenly Spheres')* more than a hundred years earlier, it would be Isaac Newton's groundbreaking *Principia* of 1687 that really set the ball rolling, scientifically speaking.

Archaeology, too, was in its formative period. Sir Thomas Browne's 1658 *Hydriotaphia* or *Urn Burial*, a study based on a cache of Anglo-Saxon burial pots discovered in Norfolk, was one of the first texts to not only describe such artefacts in detail but also to survey all that was known about funerary practices at that time. In Urn Burial Browne uses the discovery of the burial urns as a springboard for philosophical speculations about matters such as mutability and impermanence, and in its final chapter contemplates the nature of death, life, and a possible afterlife.[1]

In keeping with the dogma of the time regarding humankind's relatively short tenure on Earth, it was believed that the hand axe found at Black Mary's Hole was made of stone simply because of its maker's lack of metallurgical knowledge. It was for this same reason that objects like Neolithic flint arrowheads, which were also dug up from time to time, were not thought to be human artefacts of great antiquity. Generally, such findings were considered to be far more recent than they actually were and were often ascribed to the fairy world or regarded as examples of 'elf-shot'.

A belief in 'elf shot', or 'elf arrows', was widespread in the north of England and Scotland in the seventeenth and eighteenth centuries and was mooted especially in cases of sick cattle, which were believed to have become afflicted as a result of arrows fired by malevolent elves.[2] A fantastical explanation, certainly, but for the late seventeenth-century religious mind such lore was easier to digest than any notion of the human species stretching back more than a few thousand years. While archaeology and all that it implied verged on heresy, fairy tales were just about acceptable. There was little consideration that objects like flint arrowheads might predate the world's prescribed short life of less than six millennia. The arc of world history was short and finite; time was truncated by Christian theological orthodoxy. While many non-European cultures embraced the concept of a world that was immeasurably ancient, in the West no such notion existed, and a religiously constrained view of the world and its creation persisted until the Victorian age.

The flint tool found at Black Mary's Hole can currently be seen on the ground floor of the British Museum, where it is identified as the Gray's Inn Lane hand axe. About 15 centimetres long and sharply pointed, it looks to have been a useful instrument for butchering elephants when no metal tools were available. The elephant tooth and the axe are almost certainly contemporaneous and taken together support the view that the location where they were found was a site for butchering animals.

The hand axe, which has since been identified to be approximately 350,000 years old and date from the Palaeolithic era, became the first implement of its kind to be recorded with a published illustration. More significantly, it was also among the first to be identified as belonging to an ancient people and not merely a Roman or 'Ancient Briton' relic. It was John Conyers who argued that the axe was a human artefact and not, as many similar objects before it had been identified, an example of 'ceraunia', or thunderbolt. Unlike elf-shot, which were believed to be of supernatural origin, ceraunia were thought to be objects created by extreme natural phenomena like violent thunderstorms. Such rare items were often attributed magical properties and used as protective talismans against lightning strikes in a similar way that unexplained objects like fossil echinoids might also be employed. Conyers led the way in recognising that objects like the hand axe found at Black Mary's Hole were not only manmade but also of ancient pedigree.

The fossil collector John Woodward (1667–1728), a near contemporary of Conyers, held similar views on fossils, in particular echinoids. Rather than considering them to be 'figured stones' of supernatural origin that had landed on Earth during thunderstorms, Woodward (correctly) believed them to be fossilised sea urchins. Woodward did not divert from accepted religious orthodoxy, however, and argued that fossils provided conclusive proof for the biblical account of the Flood, and that the fossils revealed when the waters receded were evidence of antediluvian creatures that had existed before that apocalyptic event took place. Another contemporary, the polymath Robert Hooke, shared the same view and expressed this in his 1665 work *Micrographia,* arguing that the fossils found in rocks were the petrified remains of organisms that had once been alive.[3] Sir Thomas Browne had earlier expressed a similar belief in his 1646 treatise *Pseudoxia Epidemica*, a work that debunked much that was hitherto considered gospel in the field of botany, geology, geography, anatomy and zoology.

*

Displayed alongside the Gray's Inn Lane hand axe in the British Museum is another flint tool of similar size and shape, one of several found at Hoxne in Suffolk in 1797. Discovered by the antiquarian John Frere, this is considerably older than its museum neighbour and estimated to be around 400,000 years old. Like the London hand axe, it has two sharply defined cutting edges. Following his discovery of this along with other 'objects of curiosity', as he called them, at Hoxne, John Frere wrote a letter to the Society of Antiquaries in which he was sympathetic to the idea of a chronology of the Earth that challenged previous notions based purely on biblical sources. An artist was commissioned to draw the hand axe to accompany an article ('Account of Flint Weapons Discovered at Hoxne in Suffolk') that was published in the Society's journal *Archaeologia* in 1800. In his article, Frere described the hand axes as being 'evidently weapons of war, fabricated and used by a people who had not the use of metals [and are from] a very remote period indeed.'

While Frere was correct about the antiquity of the site, he was mistaken about the purpose of the tools he found, which were probably used for butchering purposes rather than anything martial. The area around Hoxne, close to the Suffolk–Norfolk border, has subsequently revealed extensive flint-working sites next to the banks of an ancient, long-vanished river. Animal bones indicate that large mammals like elephant, rhinoceros and lion used to range here during a warm interglacial period between 425,000–360,000 years ago, a span that ties in with the time when humans occupied the site and the axe was in service. The site has subsequently been considered so significant that the village name has been immortalised as a period of the middle Pleistocene: the Hoxne Interglacial.

The late eighteenth and early nineteenth centuries witnessed a flow of more enlightened thinking as advances in geology, archaeology and zoology pushed time's boundaries beyond what had previously been held to be true. The final geological proof of the antiquity of artefacts like flint hand axes came when businessman and archaeological enthusiast John Evans discovered that tools he had found at Saint-Acheul in the Somme region of France in May 1859 were almost identical to those found sixty years earlier at Hoxne. Evans's response was to immediately visit the Hoxne site with his colleague Joseph Prestwich, where he found more axes of similar design at the pit itself along with several that were already in the collections of local antiquaries. The two men then returned to the Somme in the company of Fellows of the Geological Society. These included Charles Lyell, the most influential geologist of the Victorian era, who up to that time had remained sceptical regarding any theories on the age of the human species. Lyell came away from Saint-Acheul a convert, with dozens of flint tools in his possession and a fresh outlook on such matters. In an address to the Annual Meeting of the British Association for the Advancement of Science held in Aberdeen in September 1859 he announced, 'I am fully prepared to corroborate the conclusions which have been recently laid before the Royal Society... I believe the antiquity of the Abbeville and Amiens flint instruments to be great indeed if compared to the times of history or tradition.'

The year 1859 proved eventful for science. Just two months after Lyell's Aberdeen address, Charles Darwin's *Origin of Species* was published. Meanwhile, the naturalist Alfred Russel Wallace had independently arrived at much the same conclusion based on research he had carried out in the Malay Archipelago. Darwin and Wallace, with their parallel development of evolutionary theories, went even further than the geologists, forcing the scientific world to consider not only mankind's antiquity but also a human ancestry that predated *Homo sapiens*. Understandably, such theories were highly controversial, especially by those who regarded the Old Testament as the last word on the matter. The controversy rages still in some of the more fundamentalist enclaves of the United State's southern Bible belt.

The term 'deep time' was first introduced by the American writer John McPhee in 1981 in his book ***Basin and Range***. It built upon the existing concept of geological time, which was first developed in the eighteenth century by the Scottish geologist John Hutton. In 1785 Hutton formulated a theory in which he imagined an endless cycle of rocks forming beneath the sea, each being uplifted and tilted in turn before the process of erosion formed new strata. His theory postulated that the Earth's physical landscape was never static but continuingly transforming over very long periods of time. Because of this, contrary to what had hitherto been orthodox religious belief, he argued that the Earth could not possibly be young but instead must be of almost unimaginable antiquity. Hutton formulated his theory based on what he had seen at several places in Scotland. The first of these was on the Isle of Arran, off Scotland's Ayrshire coast.

I came upon the site of what became known as Hutton's Unconformity while on holiday on Arran. Taking a walk along the coast northwest of Lochranza on the path that leads to Newton Point, I encountered a large sign that described Hutton's geological work on the island. It quoted Hutton's words regarding his visit to Arran, which was quite single-minded in its mission:

'In setting out on that expedition, I had but one object in view; this was the nature of the granite, and the connection of it with the contiguous strata.'

Hutton came to Arran in 1787 to gather evidence to prove that granite was a rock that had formed deep within the Earth, and which prior to its breaking the surface had existed in a molten state. This was at odds with the contentious 'Neptunist' theories of his peers, who considered granite to be a sedimentary rock that (like all rocks) had formed either in the sea or during the Biblical Flood.

What Hutton discovered at Newton Point proved even more revolutionary in terms of advancing geological knowledge. Here on the seashore, he observed layers of gently tilted sandstone, which, in turn, sat upon more steeply tilted metamorphic rocks. This unusual and disjointed sequence, he reasoned, must represent millions of years of geological time. As we now know, the sandstone in question is around 350 million years old, while the metamorphic Dalradian rock beneath it can be dated to be around 540 million years old. Consequently, a time gap of almost 200 million years separates the two adjoining strata. This was the first of several 'unconformities' –locations where there is a break in the sequence of sedimentary deposits and where newer rock lays on top of older – that Hutton found in Scotland.

During a boat trip around Siccar Point in Berwickshire the following year, two of Hutton's friends and colleagues, mathematician John Playfair and fellow geologist James Hall, also became convinced of Hutton's theory. The example of unconformity so clearly on show at Siccar Point is that of gently sloping beds of late Devonian red sandstone on top of near vertical beds of greywacke from the early Silurian age. The sight of this juxtaposition of geological ages was so mind-blowing to Hutton's accomplices that it prompted Playfair to exclaim that it was like 'looking so far into the abyss of time'.

Sites like Siccar Point would go on to inspire later geologists like Charles Lyell, who developed Hutton's ideas further to come up with the uniformitarian paradigm upon which all modern geology is founded. Lyell, in turn, would have considerable influence on Charles Darwin. In the wake of the pioneering work of the likes of Conyers, Hall and Frere, antiquarianism evolved from its amateur status to become the academic discipline of archaeology. Following Lyell and Hutton, geology went on to become a branch of natural science. Both disciplines embrace the distant past as they explore the depths of what has since become known as deep time.

The notion of deep time carries more philosophical meaning than straightforward geologic time. Robert Macfarlane, in *Underland*, puts it simply, describing deep time as 'the chronology of the underland'. He writes, 'Deep time is kept by stone, ice, stalactites, seabed sediments and the drift of tectonic plates,' before adding, 'there is a dangerous comfort to be drawn from deep time'.[4] While recognition of the transitory nature of all things, however solid and unmoving they may appear, might be considered 'a dangerous comfort', in the existential gloom of a worsening global crisis some might view such an acceptance as little more than cold comfort.

The concept of deep time is frequently cited in the context of raising awareness of humankind's place in nature and the fragility of the human-wounded Earth. Deep time takes into account the vast tract of time that existed before *Homo sapiens* – or indeed, any hominin – evolved, and how humans might even be considered to be almost an afterthought, an irrelevance. An awareness of deep time gives us a sense of place (and time) in the scheme of things. It shows us the incomprehensible greatness of the universe and the microscopic smallness of our place in it. On a good day it might even give us a sense of humility.

*

Sometime around the middle of the twentieth century the Earth entered an age when the Holocene period, which began at the end of the last ice age around 12,000 years ago, gave way to the Anthropocene, the 'age of humans'. The Anthropocene can best be defined as that epoch in which human activity has had a dominant effect on the world around us, on its climate, environment and ecosystems. The precise date of its onset remains a matter of debate. James Lovelock, originator of the *Gaia* concept, claims that the Anthropocene started with the Industrial Revolution in the early nineteenth century, the period in modern history when the use of fossil fuels for manufacturing and transport got fully underway. Fine-tuning this connection between the dominance of human influence and technological progress, it could even be said that the Anthropocene period began with the invention of Thomas Newcomen's steam-powered pump, a machine first used to remove water from a coal mine near Dudley in the English Black Country in 1712, and which made coal mining viable on an industrial scale.[5] While this time-frame is widely accepted, it is equally possible to hold the view that the onset of the Anthropocene can be pushed further back in time to the dawn of agriculture, to the Neolithic period when the greenhouse gas methane started increasing in concentration in the atmosphere as a result of the domestication of cattle and the planting of rice paddies in eastern Asia.

Some institutions like the International Union of Geological Sciences still contend that the Holocene and Anthropocene are one and the same in terms of geological time periods, but in 2016 an interdisciplinary research body called the Anthropocene Working Group established that the two were different and that the Anthropocene began in 1950 when the Great Acceleration, the dramatic surge in human activity affecting the planet that had begun with the Industrial Revolution, started to curve steeply upwards.[6] Since World War II, technology, world population and consumption have all been in the ascendant and have continued to increase in an accelerated manner as pollution, scarcity of natural resources, irreversible climate change, extinction or loss of many species, the threat of famine, drought and ever more virulent pandemics all vie for membership of the quartet of apocalyptic equestrians predicted in the *Book of Revelation*. Unceasing global warfare, with its catastrophic human and environmental consequences, is, of course, almost certainly guaranteed a place.

The dilemma of modern times seems to be that technology and scientific innovation has run away with itself. Too much, too soon: as a species we cannot keep up with ourselves. In evolutionary terms, we – humanity at large – have just not had enough time to adjust to our own capabilities. Too clever by half, we seem destined to serve the technology available, regardless of whether we need it or not. The hand serves the axe, the tool seeks the operator. Electronic gewgaws like mobile phones and laptop computers become obsolete almost as soon as they leave the mass distribution depots. There will always be a newer version: 2.0, 2.1 – fresh objects of desire. Meanwhile, even here in the United Kingdom, the world's sixth-richest country (or ninth, depending on which measures of GDP are used), we have food banks, homelessness and children going hungry.

Despite the pronouncements of the Anthropocene Working Group, another line of thinking might place the precise starting point of the Anthropocene to the time of the first atomic bomb tests and the devastating terror of their consummation at the Japanese cities of Hiroshima and Nagasaki in 1945. Thermonuclear testing continued apace in the years that followed, casting a dark shadow over any possibility of worldwide harmony as the Armageddon threat of the split atom was exploited by conflicting ideologies and economic systems: a global Cold War ghosting the tangible fear of what might be a white-hot conclusion. The argument has always been, of course, that a mutually held fear of such terrible consequences would provide sufficient incentive to maintain a fragile peace.

Geology and nuclear physics are uneasy bedfellows in the layers of deep time that stretch away to eternity. The half-life of depleted uranium (U-238) is roughly the same as the age of the Earth – about 4.5 billion years. The half-life of C-14, the isotope used for carbon dating, is around 5,730 years – a timespan that stretches back to the early Neolithic in the British Isles, and roughly the same age that Christians believed the Earth to be back in the seventeenth century. Meanwhile, the Anthropocene is perhaps seventy-five years old and counting.

*

Modern scientific thinking has pushed back the age of the Earth from around 6,000 years to a figure that is now over 4.5 billion years, an unimaginable span of time. The Universe itself is three times older. Our most advanced radio telescopes are capable of detecting the ghostly whispers of events that took place close to the beginning of time: event horizons where the existing laws of physics are stretched beyond comprehension, where energy-radiating quasars powered by super-massive black holes swallow stars and collapse gravity.

The vast expanse of years that marks the age of the planet holds the geological era we know as the Cretaceous. Sandwiched between the Jurassic and Cenozoic eras, this was a time of shallow warm seas teeming with marine life: microscopic organisms that would be preserved in perpetuity as chalk, larger creatures that would endure as fossils; sponges that under pressure would dissolve into liquid silica then slowly harden to form nodules of flint. Even this span of time – around 80 million years – is unthinkable in human terms, and it does not help to note that it took place at a time much closer to the present than it was to the birth of the planet. Measured on the same timescale, the arrival of post-glacial Quaternary flint in East Anglia is something that floats far closer to the surface of the well of deep time. It is the most recent blink of an eye, a five-to-midnight spasm of shape-shifting of a territory that had once been roamed by an earlier version of what we would eventually become. Our distant *Homo antecessor* predecessors, the 'Pioneer Man' hunter-gatherers who left behind footprints and flint axes at what is now Happisburgh, Norfolk, would have experienced a topography quite different to that of today.

We measure life by events, by the parallel lives of family, friends, neighbours and loved ones. We measure it by breaths and birthdays, days and sleeps, or even, as T. S. Eliot asserted, in coffee spoons. The scale of time is hard to grasp beyond the framework of our own lifetime and that of our immediate ancestors. The three dimensions of distance are easier to understand. The fact that our satellite, the Moon, is on average a little less than a quarter of a million miles away is comprehensible in human terms (that equates to two and a half times 'round the clock' in an old but dependable motor vehicle). The knowledge that the Moon's light takes just over a second to reach us on Earth is more difficult to comprehend, especially when it is remembered that this is reflected light from the Sun, which lies 93 million miles away and whose light takes around eight minutes and twenty seconds to reach both Moon and Earth. Light, we instinctively feel, is something that is instant and simultaneous. It is easy to overlook that it, too, sometimes has far to travel.

*

As a child I was fascinated by astronomy and space travel. Inspired by Patrick Moore's *The Sky at Night* on the television and textbooks borrowed from the library, I became for a while quite an earnest scholar despite spending much of my time at school staring absent-mindedly out of the window. In truth, my mind was not so much absent as thinking of other worlds. I was enthralled by the distances involved and the sheer immensity of space. I could just about get my young head around the vastness of the Solar System with its increasingly distant, orbiting planets that circuited the star we shared at its centre. But beyond this, the quantum leap of seemingly impossible distances across outer space, and corresponding leap of imagination required to imagine galaxies and stars that had died before our own planet was even born, made my head spin. Immense distance enthralled me, as did the notion of deep space. The notion of deep time, though, was altogether stranger and even more difficult to grasp.

Flint, born in the warm seas of the Late Cretaceous, a time when dinosaurs like *Tyrannosaurus rex* still stalked the land, provides a material connection with deep time. It gives a point of reference for the geological past, a way marker for shared human experience. It connects our own brief lives with those who went before. We measure the time since flint-knapped churches were first erected in the flatlands of East Anglia and the chalk country of southern England in hundreds of years, in dozens of generations. Think of the Roman presence in these isles, a time of road- and fortress-building, and the span enlarges to two thousand years before the present; perhaps eighty generations ago. Triple or quadruple this and we reach back to the Mesolithic period when bands of hunter-gatherers braved the cold to venture across the land bridge from Europe. Killing game with flint-pointed weapons, carving up meat, scraping skins and making fire with the same treasured stone, these ancestors who strayed across Doggerland hundreds of generations before us trod much the same ground as we do today, although it is doubtful that we would recognise it.

Homo sapiens emerged around 300,000 years ago but, long before this, early hominins were also tool users who exploited flint to suit their needs. The rock from which the tools were made came into existence long before this, in shallow warm seas around the time of dinosaurs. This is not the sort of chronology that can be easily plotted on a linear scale, but all of it – the Cretaceous formation, the Palaeolithic axe-making, the medieval church building – can be reckoned as points along the same temporal line: a line of destiny that we might call flint time.

Flint connects human evolution, scientific advance and cultural progression. It cuts, sparks, supports and beautifies. It makes a tool, an ornament, a wall for houses and temples, a road for feet, a strike-a-light for fire. Since its first use by long-extinct hominin ancestors, flint has set in motion a slowly accelerating chain reaction of discovery and exploitation of the world around us. Even today, as the denouement of the digital revolution that has succeeded industrial-scale manufacturing plays out, the iconic form of a flint axe continues to hold deep symbolic meaning. Even London's Shard, the tallest building in Western Europe, a cloud-piercing totem to the power of City capital (and Qatari wealth), adopts the form of a dagger created from fractured stone. There is a clue in the name. It might equally be called the Flake.

The flint axe represents, as Alan Garner puts it, 'our first meddling with the Universe by intelligence'. Referring to the Lowell Telescope at Jodrell Bank close to his home in Cheshire, he writes, 'The pebble opens the Cosmos. If mind had not interfered with pebble, if hand had not struck flake, no dish would hear, and no antenna focus.' Garner also notes that the subject of his discussion, a half-million-year-old flint axe found in a nearby field, is 'of Abbevillean execution', wryly concluding with the question, '...will the flint that shaped us be our nemesis, our nemesis of Abbevillean execution? We've been here before... Abbeville is in the valley of the Somme.'[7]

There are echoes of this same thinking in Garner's fiction too. In *Boneland*, the long-awaited conclusion to a trilogy that began half a century earlier with *The Weird of Brisingamen*, an Achuelian axe, extracted from the earth beneath the most universe-penetrating of the radio telescope dishes at Jodrell Bank, is central to the story. The axe – 'Abbevillean. Or Acheulian. Non-derived. Pre-Anglian. MIS 13' – is now a paperweight, 'a comforting stone to hold'. It is the black stone left by the story's other protagonist, a solitary Palaeolithic figure who may or may not be Neanderthal. The stone, 'the little thing that held the star', survives underground until its rediscovery in the modern age. The stone axe is the link that holds the various threads of the story together, connecting the deep past with deep space at a location where the universe is laid bare as the death throes of even the oldest stars are monitored. Time is fragmented, layered and multi-dimensional, yet as Colin, the brilliant astrophysicist in *Boneland* asserts, as humans and scientists, 'we've opted for the advancing linear flow' as it is the most efficient for our needs. It is all a matter of perspective.[8]

Our rock-shifting, flint-knapping ancestors looked at the world through eyes that were very different to ours. Almost certainly, they also held a different view on the passing of time. For them time was not the linear trajectory of a life from birth to death but something more cyclical in nature. Adam Thorpe in *Silbury Hill*, the author's multi-layered meditation on the prehistoric structure that has figured in his life since he lived in its shadow as an unhappy boarder at Marlborough School, considers how the hill's Bronze Age builders would have viewed the passing of time. 'There might never have been a want of the stuff (time), because it was a continual roll unnoticed by clocks, on which the past was visible not as something recorded (old photos, old manuscripts) but *as something that hung around, part of the same material as the present*, a touch ghostly but just as potent as living flesh.'[9] A past that hung around, part of the same material as the present; part too, by extension, of the same material that would become the future. Thorpe continues to explain how each stage of the ongoing Silbury project represents an 'eternal now' within the living monument, and how the hill embodies the movement of ever-renewing cyclical time. He describes such an understanding as essential to those who lived in the distant past, as a 'great saving wisdom in which death is manageable. Otherwise to be conscious of mortality is unbearable, an unbearable thing'.

Exchange locations and times, swap Silbury Hill for Stonehenge, or Grime's Graves or Cissbury Ring, and the perception is unchanged: prehistoric time flowed cyclically, like the shift of the seasons, like the moon; like the tide, lapping lives in its dependable rhythm. For the monument builders of the Neolithic we can only assume that the linear flow of time held little importance in regard to multi-generational projects like the stone-raising of Stonehenge, or the colossal amount of earth-shifting involved in constructing Silbury Hill. Such Sisyphean tasks would have spanned multiple generations and the view of the world perceived by the great grandchildren of those who initiated the project could have been little different to that of their forefathers. The stones were a constant presence; the artificial mound always loomed large. The purpose was to maintain and enlarge those manmade structures that, as they perceived it, had always been there. If Stonehenge and other monuments functioned partly as calendars, as the archaeological evidence seems to suggest, then they were monuments that marked time's annual cycles. Life – human, animal, spiritual – was not a trajectory but a circle.

Time is different in the modern world. Its clocks run differently. The pace is such that the past, present and future are becoming conflated. As a society, even if not as individuals, we are often only concerned with the near future. We are obsessed with the immediate, the very soon; with tomorrow. Nothing asserts this view more boldly than the current threat of climate change and its potential environmental, political and social impact on an already deeply troubled biosphere. The truth is we live on a planet that is held hostage by its most advanced life form. We feel powerless against others of our own species. Perhaps a safe, stable future has become so unimaginable that we turn our gaze inwards, ignoring the horrors that loom so close on the horizon. As has already been noted, it is easier to imagine an end to the world than an end to capitalism.[10]

We talk incessantly, in clichés, of our children's children's future, yet it is hard to even imagine what their lives might be like in material terms. All we can predict is that it will be a more polarised world of rich and poor, haves and have-nots. Lives that will be either perilously precarious or snugly secure, preoccupied with either a relentless search for food, water, shelter and fuel or with the blotting out of bad news by means of chemical or electronic escapism.

The future draws closer, converging ever nearer to the present but never quite touching. Whatever fate unfolds in the years to come, deep down in the chalk, flint will still be there – layered, glistening, silent but replete with the stone memory of deep time. Unlike fossil fuels and precious metals it will not be ruthlessly exploited and dragged from the earth until no trace remains. Flint has outlived its usefulness; its place in the arc of human history is guaranteed, but as a resource it is no longer vital. In modern retail parlance, flint is a discontinued line. Silicon may still be central to our modern lives but it takes the form of microchips, not microliths.

No longer essential for our needs, flint signifies a redundant technology and an archaic way of doing things. Even so, unless something truly cataclysmic takes place, those medieval churches with their finely wrought flushwork will still be standing centuries from now as repositories of history. Mesolithic blades and scrapers will continue to turn up on beaches. Roughly-hewn flint blocks, gathered and worked centuries ago – palimpsests recycled from earlier structures – will still adorn rustic walls. On shingle spits, flint's harsh music will still be there for those with ears to listen. Or at least, the ghost of its song: its tinkle and ping, its mineral grating. We will continue to walk on trails that have flint at their core; paths that have seen generations of feet venture further and further in the span of a lifetime.

And beyond our time, when the Anthropocene eventually shudders to an end, the rocky foundations of this world will stay firm. Beneath the surface, flint will lie cradled in the chalk, unmoved by ice or industry. There, gleaming, unsullied by air or daylight, as black as an empty void, will sit a jewel forged by warm seas and primordial dust: a wonder that is full of stars.

ACKNOWLEDGEMENTS

Short extracts of this book have appeared in different forms in *Echtrai Journal*, *hidden europe* magazine and *Stravaig*, the online journal of the Scottish Centre for Geopoetics. Thanks go to Baz Nicholson, Nicky Gardner, Susanne Kries and Norman Bissell for considering them worthy of inclusion.

I am grateful to Sara Hunt and her team for taking interest in and publishing this, my second title for Saraband. I am also indebted to Professor David Howe, who was good enough to discuss the outline of this book with me at an early planning stage. Sincere thanks go to the following friends for their company on some of the journeys mentioned in the book. In Norfolk: Jeremy Salmon and Christine Isaacs at West Runton beach; Hanne Siebers and Klausbernd Vollmar at Blakeney Point, and John Hiskett for venturing underground with me in Norwich. In Suffolk: Clare Brennan and Martin Daly, who crossed the river with me to Orford Ness. In London: Nigel Roberts for the walk to King's Cross, site of the Gray's Inn Lane hand-axe, by way of the hidden River Fleet. In Northern Ireland, thanks go to Chris and Jen Cordwell for time shared in Belfast and along the lovely Antrim coast. It goes without saying that my wife Jackie, as well as being wholly supportive throughout, was also a willing and enthusiastic accomplice on most of the above, as well as on numerous other flint-themed outings that go unmentioned here.

My gratitude goes to Dr James Dilley for his patient instruction on flint knapping, Hugh Lupton for his storytelling and Peddars Way myth-making, and Tim Holt-Wilson for sharing a geological knowledge that extends from the quaternary deposits beneath our feet to glacial erratics in medieval church walls. My thanks also extend to Ruth Brumby at Cromer Artspace for the discussion on flint in art, and to Yelena Popova for generously sharing her artistic vision and for the gift of the flint 'scholar stone' that sits on my desk.

ENDNOTES

One: Flintlocked

1. *Norfolk Churches* website, www.norfolkchurches.co.uk, Simon Knott
2. *Ancestors,* Prof Alice Roberts

Two: Flint Eastwards

1. *The Norfolk Landscape,* David Dymond
2. ibid
3. *The Pebbles on the Beach,* Clarence Ellis
4. *Stones of Aran: Pilgrimage,* Tim Robinson
5. *Through England on a Side Saddle in the Time of William and Mary,* Celia Fiennes
6. *Norwich: Archaeology of a Fine City,* Brian Ayers
7. www.norwich360.com/guildhall.html
8. https://britishlistedbuildings.co.uk
9. *Ancient Wonderings,* James Canton
10. *Highways & Byways in East Anglia,* William A. Dutt
11. *The Roman Camp and the Irish Saint at Burgh Castle,* Louis H. Dahl
12. https://www.heritage.norfolk.gov.uk NHER number 10500
13. https://www.heritage.norfolk.gov.uk NHER number 12776, 9605, 9592
14. https://www.heritage.norfolk.gov.uk NHER number 6100

Three: On the Beach

1. *Flint: Its Origin, Properties & Usage,* Walter Shepherd
2. *Flint Architecture of East Anglia,* Stephen Hart
3. 'Flint', Alan Garner, *Cornerstones,* ed Mark Smalley
4. https://www.nhm.ac.uk/discover/how-an-asteroid-caused-extinction-of-dinosaurs.html
5. *The Day Before Yesterday: Five Million Years of Human History,* Colin Tudge
6. https://en.wikipedia.org/wiki/Flint
7. *Mining and Quarrying in Northern Europe: A Social Perspective,* ed. Anne Teather, Peter Topping, Jon Baczkowski
8. *An Illustrated Dictionary of the Gods and Symbols of Ancient Mexico and the Maya,* Mary Miller, Karl Taube
9. *Antiquity magazine* 7/10/2021 published online by Cambridge University Press https://www.cambridge.org/core/journals/antiquity

10. *The East Country: Almanac Tales of Valley and Shore,* Jules Pretty
11. *The West Runton Elephant (Cromer Museum brief history guide No 17),* Nigel Larkin
12. http://www.northfolk.org.uk/geology/west%20runton-east%20 runton
13. 'Returning to the source: investigating flaked stone tool raw material procurement strategies in prehistoric Ireland', https://www.lithicsireland.ie/archaeology_projects
14. 'Star of the Sea: on urchins and ancient waters' (in *Elementum* edition four), Alex Woodcock
15. 'Country diary: village church with unusual adornment', Nicola Chester, *Guardian,* 6/3/2020
16. *The Star-Crossed Stone: The Secret Life, Myths, and History of a Fascinating Fossil,* Kenneth J. McNamara
17. *Shepherds' crowns, fairy loaves and thunderstones: the mythology of fossil echinoids in England,* Kenneth J. McNamara, Geological Society, London Special Publications
18. 'Fossil folklore: Echinoderms', Dr Paul D. Taylor, *Deposits Mag* 04/04/2017 https://depositsmag.com
19. *Mudlarking: Lost and Found on the River Thames,* Lara Maiklem
20. *Thunderstone,* Nancy Campbell
21. *Dragons' Teeth and Thunderstones: The Quest for the Meaning of Fossils,* Ken McNamara
22. *Sir Gawain and the Green Knight,* translated by Simon Armitage
23. *The Engines of Our Ingenuity,* John H. Lienhard, https://www.uh.edu/engines/epi1243.htm
24. *The Garden of Cyrus,* Sir Thomas Browne
25. https://ukfossils.co.uk/2008/01/21/east-runton

Four: Flint Music

1. *Ghostland,* Edward Parnell
2. https://www.ancient-origins.net/history-famous-people/boudicca
3. *The Rings of Saturn,* W. G. Sebald
4. *I See a Silence,* Ilya Kaminsky
5. 'Artangel: Afterness review – an island of secrets that's its own work of art', Laura Cumming, *The Guardian* 27/06/2021
6. *Ness,* Robert Macfarlane & Stanley Donwood
7. *Independent Research Group Orford Ness,* https://www.irgon.org.uk/laboratory-1
8. *The Rings of Saturn,* W. G. Sebald
9. *The Bodies on the Beach: Sealion, Shingle Street and the Burning Sea Myth of 1940,* James Hayward
10. *The Shingle Street Shell Line,* Els Bottema and Lida Kindersley

11. *Flint: Its Origin, Properties & Uses*, Walter Shepherd
12. *1989 Annual Report of Rye Harbour Nature Reserve*, Michael Alford

Five: Caught Knapping

1. *Ancient Craft* website, www.ancientcraft.co.uk
2. *Lapidarium: The Secret Lives of Stones*, Hettie Judah
3. *Prehistoric Flintwork*, Chris Butler
4. *Kindred: Neanderthal Life, Love, Death and Art*, Rebecca Wragg Sykes
5. *Science magazine, 1/05/2019*, Ann Gibbons www.science.org
6. *The Inheritors*, William Golding
7. *Kindred: Neanderthal Life, Love, Death and Art*, Rebecca Wragg Sykes
8. *The Brecks Earth Heritage Trail*, Tim Holt-Wilson
9. *Norfolk Churches* website www.norfolkchurches.co.uk Simon Knott
10. https://canmore.org.uk/site/21303/den-of-boddam
11. 'Swine team: pigs help uncover ice age tools on Scottish island', *The Guardian*, 9/10/15, Maev Kennedy
12. *Storakaig and Rubha Port an t-Seilcih Archaeological evaluation of new Mesolithic sites on Islay*, Steven Mithen and Karen Wicks
13. *Land of the Ilich: Journeys into Islay's Past*, Steven Mithen
14. *Current Archaeology*, June 2010
15. *The sources of flint and chert in northern Britain* (*Proceedings of the Geological Society 1977—8*), C. W. Wickham-Jones and G. H. Collins
16. *Prehistoric Flintwork*, Chris Butler
17. *Grounded*, James Canton
18. *Prehistoric Flintwork*, Chris Butler

Six: Rock of Ages

1. *Grime's Graves English Heritage Guidebook*, Peter Topping
2. ibid
3. *Prehistoric Flintwork*, Chris Butler
4. ibid
5. https://the-past.com/review/whats-on/john-pull-worthings-hero-archaeologist/
6. *The Pattern Under the Plough*, George Ewart Evans
7. ibid
8. ibid

Seven: Game of Stones

1. *A Norfolk Songline: Walking the Peddars Way*, Hugh Lupton & Liz McGowan
2. *Grime's Graves English Heritage Guidebook*, Peter Topping
3. *Water and fire as transformation elements in ritual deposits of the Scandinavian Neolithic* (Lund University), Lars Larsson

4. *Axe makers of the Wahgi: Pre-colonial Industrialists of the Papua New Guinea Highlands,* John Burton
5. *Seahenge: An archaeological conundrum* (English Heritage), Charlie Watson
6. *The Tale of an Axe: How the Neolithic Revolution transformed Britain,* David Miles
7. ibid
8. https://salisburymuseum.org.uk/collections/stonehenge-prehistory/jadeite-axehead
9. https://www.bbc.co.uk/ahistoryoftheworld/objects
10. *The Tale of an Axe: How the Neolithic Revolution transformed Britain,* David Miles
11. *Ancestors,* Alice Roberts
12. *Rathlin: Nature and Folklore,* Philip Watson

Eight: Faith in Stone

1. *Tour through the Whole Island of Great Britain* (1725), Daniel Defoe
2. www.suffolkchurches.co.uk, Simon Knott
3. *Landscapes of Detectorists,* ed. Innes M. Keighren & Joanne Norcup
4. https://www.roundtowerchurches.net/
5. https://www.roundtowers.org.uk
6. *Flint Architecture in East Anglia,* Stephen Hart
7. ibid
8. https://www.roundtowers.org.uk
9. https://britishlistedbuildings.co.uk/101032028-holy-trinity-church-barsham
10. https://fairsarchive.org
11. *These Feral Lands Volume 1,* Laura Cannel, Stewart Lee, Kate Ellis, Polly Wright, Jennifer Lucy Allen
12. https://www.becclesandbungayjournal.co.uk 27/02/2009

Nine: Flushed with Pride

1. *Flint Architecture in East Anglia,* Stephen Hart
2. ibid
3. *England's Thousand Best Churches,* Simon Jenkins
4. www.norfolkchurches.co.uk, Simon Knott
5. *England's Thousand Best Churches,* Simon Jenkins
6. *Building in England down to 1540: A Documentary History,* L. F. Salzman
7. www.suffolkchurches.co.uk, Simon Knott
8. 'Archaeologists map lost medieval Suffolk town of Dunwich under the sea', *The Guardian,* 10/5/2013

Ten: Holding Fire

1. http://www.breakingnewground.org.uk/earthheritagetrail/beeches-pit/
2. *Kindred: Neanderthal Life, Love, Death & Art,* Rebecca Wragg Sykes
3. *Flint: its origin, Properties & Uses,* Walter Shepherd
4. *Ways of Seeing,* John Berger
5. *Flint—The Versatile Stone,* H. J. Mason
6. *A Gun-flint Industry at Martins Clump, Over Wallop, Hampshire,* Martin J. Fowler
7. *Flint—The Versatile Stone,* H. J. Mason
8. *Breckland,* Olive Cook
9. www.brecksoc.org.uk
10. www.historicalfirearms.info
11. *Mudlarking: Lost and Found on the River Thames,* Lara Maiklem
12. *In Defence of Landscape: An Archaeology of Porton Down,* David Ride

Eleven: Chthonic Boom

1. 'Climate and Cult,' J. Walter Fewkes, a paper published in the *Report of the Eighth International Geographic Congress* 1904. Referred to in Murdo Eason's blog *From Hill to Sea* (https://fromhilltosea.com) 2014/10/30
2. *Subterranean Norwich: The grain of the city,* Matthew Williams
3. *A Tour Through the Whole Island of Great Britain* (1724-26), Daniel Defoe
4. ibid
5. *Subterranean Norwich: The grain of the city,* Matthew Williams
6. ibid
7. ibid
8. ibid
9. ibid
10. *Eastern Evening News,* May 11, 1936
11. *The Gift of Stones,* Jim Crace

Twelve: Nest of Stones, Nest of Artists

1. https://www.metmuseum.org/art/collection/search/557953
2. https://colonelunthanksnorwich.com/2019/09/15/the-norwich-school-of-painters/
3. https://inexpensiveprogress.com/2837/happisburgh/
4. *Drawn from Life,* John Skeaping, 1977
5. *Henry Moore Remembered: The Collection at the Art Gallery of Ontario in Toronto,* 1987, Alan G Wilkinson

6. *Nature and Inspiration: Henry Moore at Houghton Hall*, 2019, Sebastiano Barassi
7. Henry Moore Foundation
8. Wilkinson ibid
9. https://www.theguardian.com/artanddesign/2014/aug/08/henry-moore-bronze-sculpture-kings-cross-station
10. *Eastern Daily Press*, 15 December 2022
11. BBC News Norfolk, 22 April 2017, https://www.bbc.co.uk/news/uk-england-norfolk-39680001
12. https://www.sainsburycentre.ac.uk/art-and-objects/l-114-man-of-stones/
13. https://www.billjackson.photography/man-of-stones
14. Rachel Whiteread biography on National Gallery of Art website https://www.nga.gov/collection/artist-info.28614.html
15. *The Poetics of Space*, Gaston Bachelard
16. *Yelena Popova: Ode to a Flint*, Norwich Castle Museum & Art Gallery booklet
17. www.yelenapopova.co.uk

Thirteen: Contemplating Dust

1. *In Ruins*, Christopher Woodward
2. *John Piper: The Forties*, David Fraser Jenkins
3. *Pleasure of Ruins*, Rose Macaulay
4. *The Icknield Way*, Edward Thomas
5. *The Age of Reason*, Thomas Paine
6. *Companion Guide to East Anglia*, John Seymour

Fourteen: Flint Time

1. *Hydriotaphia or Urn Burial*, Sir Thomas Browne
2. *Getting Shot of Elves: Healing, Witchcraft and Fairies in the Scottish Witchcraft Trials in Folklore* (2005), Alaric Hall
3. *Dragons' Teeth and Thunderstones: The Quest for the Meaning of Fossils*, Ken McNamara
4. *Underland*, Robert Macfarlane
5. *The Tale of an Axe: How the Neolithic Revolution transformed Britain*, David Miles
6. https://education.nationalgeographic.org/resource/anthropocene
7. *Flint*, Alan Garner, Cornerstones – Subterranean Writings, ed. Mark Smalley
8. *Boneland*, Alan Garner
9. *On Silbury Hill*, Adam Thorpe
10. *Capital Realism*, Mark Fisher

INDEX

Index